MW01484457

PSALMS
IN MY
BACKPACK

Walk with the Lord!
Linda Jane Niedfeldt
— Janie

PSALMS IN MY BACKPACK

154 Vignettes from Our Appalachian Trail Hike

Four Kids, One Husband, and Me

LINDA JANE NIEDFELDT

XULON PRESS

Xulon Press
2301 Lucien Way #415
Maitland, FL 32751
407.339.4217
www.xulonpress.com

© 2020 by Linda Jane Niedfeldt

All rights reserved solely by the author. The author guarantees all contents are original and do not infringe upon the legal rights of any other person or work. No part of this book may be reproduced in any form without the permission of the author. The views expressed in this book are not necessarily those of the publisher.

Unless otherwise indicated, Scripture quotations taken from the English Standard Version (ESV). Copyright © 2001 by Crossway, a publishing ministry of Good News Publishers or from The Holy Bible, Evangelical Heritage Version. Copyright 2019 Wartburg Project, Inc. Northwestern Publishing House. Used by permission. All rights reserved.

Printed in the United States of America.

ISBN-13: 978-1-6312-9033-6

Acknowledgements

Thanks,

-to Tom Niedfeldt for encouragement on the Appalachian Trail (AT) and during life's hikes, for sharing his journal entries, for shooting the original photos, for drawing the map, and for being an enthusiastic reader

- to Carla Jahnke of Carla Jane Photography for family and author photos, for expertise in photo editing and in electronic transfers, and for loving the first, terrible draft of this memoir

-to Caleb Niedfeldt for his perceptive pencil drawings of life on the AT, sketched at age fourteen

-to non-family readers for their comments and guidance: Ric Alexander (AT 1990-Trail Name: Total Recs), Jean Guedes, Carol Kolosovsky, Kenn Kremer (reader/editor), Carol Stuebs, Pastor Michael Weigand

-to Liz Tolsma who not only edits, but gives kind support

-to all the hikers and friends we met along the trail, to those whose names are real and to the few whose names I changed

In peace I will both lie down and sleep; for you alone, O Lord, make me dwell in safety (Psalm 4:8 ESV).

Table of Contents

Introduction

In 1990 our family spent three weeks hiking the Appalachian Trail. It was hard, really hard. Harder than we could have imagined.

When we walked away from our car, we left the comforts of home, comforts like beds, air conditioning, microwaves, flush toilets, and even roofs. We didn't have cell phones, because nobody had cell phones, so we couldn't call for help. We couldn't check Google Maps if we became disoriented. Our only weather forecast came from studying the sky.

We had to learn how to deal with danger and discomfort, how to make the best of the worst situations, and how to even laugh at them. At times, my personality roller coasted— loving my life one minute and being filled with self-pity the next.

But along our difficult trek, we learned that we can claim the Psalms of praise and petition. King David and the other writers of the Psalms lived in a rugged world, sometimes behaving in bizarre ways, fluctuating from happy to miserable. Like them, we squatted over a fire to cook or stay warm. Like them, we were afraid of storms and wild animals. Like them, we tramped long, rocky distances and forded streams, endured the sun and cold and even slept on the ground. Like them, we praised God for his beautiful creation and steady protection.

In this memoir, written in vignette style, a Psalm verse precedes each vignette. I chose each verse because I think it applies to the situation. As you read, perhaps you'll agree with me. You may even discover connections between the Psalms in my story and your own life experiences. I'd like to encourage you to keep a parallel journal, applying the Psalms that I used in my memoir to your own story.

How It Started

I can still see the inspiring photo in a *Readers' Digest* story: a lone Appalachian Trail (AT) hiker is strolling across the Hudson River on Bear Mountain Bridge. Peaceful mountains and rolling hills are the backdrop.

"Tom, doesn't this look interesting?" I showed him the photo. "You think our family could hike some of the AT?"

Or was it Tom that called me over. "Look at this article. Hiking the AT would be a great adventure for our family."

We're not sure who had the idea. Some days we both claimed it. Other days we both blamed our spouse. But we pursued and expanded the idea.

Tom and I hoped this hike would build memories for us as a family and bond us all. We expected to emerge at the other end in great physical, emotional, and spiritual shape.

We dove into a year of planning: setting the hiking schedule, purchasing supplies, dehydrating food, packing boxes with week-long food supplies, taking a first aid course, and training physically, but never actually backpacking. We flipped through maps and books, setting the challenging goal of hiking 500 miles north from Harrisburg, Pennsylvania, to Hanover, New Hampshire, in two months.

Meet The Family

Two months to hike 500 miles was no small challenge for Tom, a forty-four-year-old Lutheran high school teacher, or for me, Janie, a thirty-nine-year-old, part-time employee and fulltime mom and wife. But to complement or complicate the endeavor, we also dragged our four children, ages nine to sixteen, into the adventure. The kids had lower expectations.

Ben, almost sixteen, was worried about starving to death, neglecting his weight-lifting program, and delaying his driver's license.

Caleb, fourteen, hated, absolutely hated, that he'd miss the summer baseball season and basketball camp. And worst of all, he'd miss out on fishing for two months.

Joel, twelve, lamented that he'd have to give up two paper routes during the warm months of summer, but he thought the hike sounded cool.

Carla, nine, was worry-free, looking forward to sending post cards to her friends and hoping to see a bear.

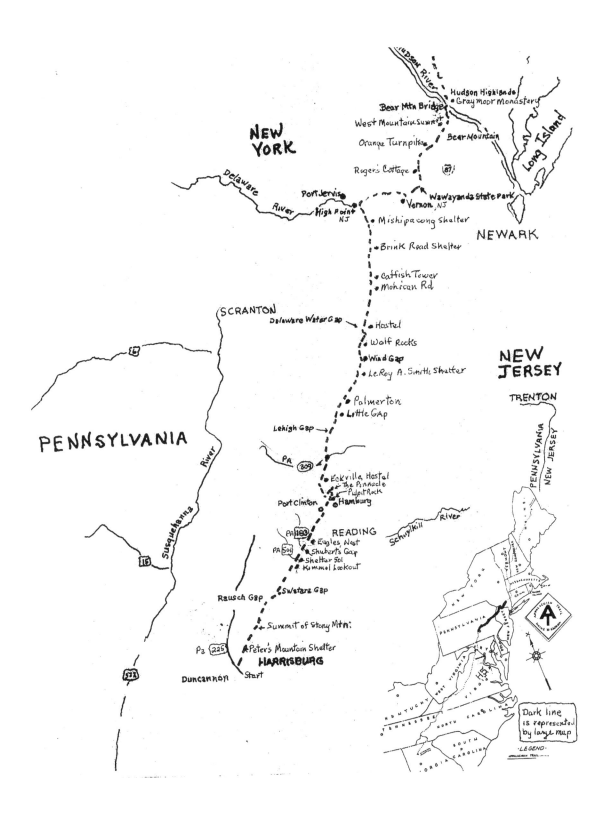

Vignette One: Surprises
Thursday, June 7, Day One

How blessed is everyone whose strength is found in you. The highways to Jerusalem are in their hearts *(Psalm 84:5 EHV)*.

"Is everyone ready?" Tom forced his enthusiasm. The rest of us stared after our car, our only means of escape, as it disappeared around a curve on the mountain highway.

I studied the empty road, paused to regain composure, and then smiled at the kids. "Of course we're ready. It will be a short walk to Peter's Mountain Shelter. How hard can 3.1 miles be this afternoon?"

That was June 7, 1990, the day we said good-bye to our car and hello to the Appalachian Trail (AT). We were near Harrisburg, Pennsylvania, where the AT crosses Pennsylvania Route 225. To our north, the AT plunged and climbed 1032.6 miles to Mt. Katahdin in Maine and to our south it also plunged and climbed 1145.7 miles to Springer Mountain, Georgia.

An acquaintance, who was storing our car at his house thirty miles away, had climbed into the driver's seat, given a friendly wave, and zoomed off. We were alone in the quiet with our packs on our backs.

And so we stepped out with excitement and trepidation. We had never done any backpacking, not even a trial run. Why shouldn't we be excited and scared? Why wouldn't we also have surprises?

"Okay, let's watch for white rectangles on the trees or rocks," Tom said. "They are called blazes, and they will guide us along the entire Appalachian Trail."

"There's one!" Carla pointed to a white rectangle, about two by eight inches, painted on a tree ahead.

"Yep, that's it," I said. "We also need to remember that one blaze means keep going straight ahead. Two blazes tell us that a turn in the trail is ahead. A blue blaze will take us on a side trail."

As we clomped along, we were always within sight of a white blaze. "Wow," Joel, the statistician, exclaimed. "Aren't you surprised that there are over 2,100 miles of the AT with white blazes every one hundred feet or so?"

"That is surprising." I grinned at his analytical brain.

Several panoramic views of farmland opened up far below. We exclaimed at the beauty, but we were beginning to wither under the blazing sun. And we were weary of squinting at the brilliant rocks, strewn haphazardly on the trail, glaring back at us in that blazing sun. But we had to be careful, and so we concentrated on the trail.

"I wonder if these patches of rocks are what the guidebooks call 'the boot killers of Pennsylvania.'" I stubbed my toe on one.

"Probably." Tom tripped ahead of me. (They weren't. That surprise would come later.)

Only one-and-a-half miles into our hike, we took a pack break.

"Here, let me help you." Tom tugged my pack off and leaned it against a tree. Putting on or taking off my pack by myself was an impossible skill—at first. I turned toward the kids and found they were assisting each other too. A surprising spirit of cooperation by rivaling siblings.

I glanced at my designed-for-women pack, stuffed with thirty-five pounds of gear, food, and supplies. The phantom weight of it still hung on my shoulders. I shrugged and stretched. Perhaps that pack was going to be tougher to carry than I figured. That was a surprise. Carrying a pack, even one designed for women, would be uncomfortable, at times unbearable.

When the break was over, we helped each other wiggle into our packs. We hiked for many more miles, it seemed. Our steps were dragging, our shoulders sagging.

"I think we missed the shelter," Ben whined. "How long can two miles be?"

We dragged and sagged some more.

Another surprise. The miles on the AT are long, not in actual length, but in the time and effort they take to hike. I began to doubt we had the strength for this adventure.

Vignette Two: More Surprises
Thursday, June 7, Day One

My heart beats quickly. My strength leaves me *(Psalm 38:10 EHV)*.

Carla had taken the lead when she gave a holler and pointed to a dilapidated, three-sided log shelter. Spying a pair of tattered socks draped over a log, we crept up and peeked around the corner into the low-roofed hut. There, leaning on his pack, propped against the inside wall, a graying, thin, unshaven man peered out at us.

"Hello." He settled back and grinned as all six of us crowded around.

He was our first encounter with a real hiker, and we drilled him with questions.

"How far are you going?" Tom jumped in first.

"About 500 miles of the trail this year. I plan to do the entire trail in 500-mile segments, one each year."

Tom nodded. "What a coincidence. We're doing 500 miles too."

"Do you have a job?" I was the practical one in the family.

"Yep, I'm on a vacation from my life in California as a surveying engineer."

I raised one eyebrow. He looked more homeless than professional. Remembering his socks, drying on the log, I asked one more question. "Do you wash your socks each night? It's important to keep your socks clean to avoid blisters, right?"

He flashed me a patient grin. "Yup, that should help." His look puzzled me, but I was relieved to know the answer to blisters. At least I thought I knew.

In a few minutes, two more men, one middle-aged, one young, sauntered into camp. "My dad and I had two bits of excitement today." The young man tossed his pack down. "We chased a hissing grouse from the underbrush this morning, and this afternoon, I almost stepped on a rattlesnake a mile or two back."

"Rattlesnake?" I gasped, remembering that nine-year-old Carla had been in the lead a mile or two back. The guidebooks had betrayed me. They had reassured me that hikers rarely see a rattlesnake. One more surprise. Guidebooks can't predict snakes.

"What did you do?"

"Oh, we just pounded our walking sticks, and the critter slithered off."

"I read that if you get bitten by a snake, it's better to get to a hospital than try to treat it yourself." I had crossed snake bite kit off my packing list.

"Well, good luck getting to a hospital on this trail," the dad piped in.

Vignette Three: Surprises And No Surprise
Thursday, June 7, Day One

In peace I will both lie down and sleep; for you alone, O Lord, make me dwell in safety

(Psalm 4:8 ESV).

My heart was still pounding from the snake story when the kids volunteered to go to the spring 300 yards downhill to pump our water supply. As they crashed through the underbrush, I yelled after them. "Don't brush against any leaves. That's how you get ticks. And snake bites," I added, lamely. There was no path to the spring, only underbrush. Surprise. It's impossible to avoid brushing leaves and bushes any place on or near the AT.

Since there was no picnic table and the shelter was taken, I balanced our little stove on the ground and squatted over the kettle to cook supper. My back hurt. Still, I wanted to impress my family and give them hope for the meals to come. We had tuna helper with two cans of tuna. (Later, we would only carry one can of tuna per meal to conserve carrying weight.)

My family surrounded me, thrusting plates out. I dipped generous servings onto each plate, but even before I served myself, the kids' plates were empty. Their mouths were like vacuum cleaners. But I had a surprise treat to conclude the meal. I presented a double box of instant vanilla pudding.

The vacuum cleaners inhaled that too.

"Is that all?" Ben slumped.

The middle-aged man sauntered over, tobacco wad in his cheek, to offer some advice for my kitchen. "Get rid of the plates. Get rid of the forks and knives. All you need is a cup and a spoon—maybe not even one per person. Ya don't wanna carry any extra weight in these packs."

I nodded and studied my array of silverware, plastic plates, bowls, and cups, but complimented myself with the fact that they were all lightweight and packed them up.

Before we could relax and join the three men for conversation, our chores had to be completed. Everyone pitched in with clean up. We set up three tents. One was for our packs, one for the boys, and one for Tom, Carla, and me. We deposited the sleeping bags in the tents.

In the next few idyllic moments, we chatted with the well-seasoned hikers about life on the trail. Caleb and Ben sprawled against a tree, Joel whittled a straight branch into a walking stick, complete with a leather band through a hole in the handle, Carla rested against me on the shelter step, and Tom leaned on the door frame, studying our newfound friends. We discovered shabby-looking people with warm hearts who, like us, were fulfilling a dream of hiking the Appalachian Trail.

As the stars twinkled in the dark sky, we crawled into our tents and sighed in exhausted relief. Soft, rhythmic breathing filled the tents, but sleep did not come for me. Too many ticks pounced and snakes slithered in my mind. There was nothing in my pack to help, but even if my pack was lacking, God was with us, watching over us, in the wilderness, in flimsy tents. I relaxed. Our safety is no surprise.

~Distance hiked on day one: 3.1 miles
~Pennsylvania Route 225 to Peter's Mountain Shelter

Vignette Four: What's in a Name?
Friday, June 8, Day Two

He counts the number of the stars. He calls them all by name
(Psalm 147:4 EHV).

Why do hikers pick trail nicknames? Maybe for fun, or to describe who they are, or maybe even to hike incognito. Everybody on the AT has a trail name. They introduce themselves with it, and they sign the registries, scattered along the trail, with it. So, of course, the three interesting characters we'd shared camp with also had trail names. But we didn't. Yet.

Roadrunner, lean and fast, was the tattered hiker we'd first met. Then the F & S Boys (father & son) had hiked into camp. They also had individual nicknames. The father was Tarheel '90, from the Tarheel state, North Carolina, and his son was Hikin' Fool, because on April Fools' Day he had quit a good job to join his dad on this hike. We felt a special attachment to these three hikers, and as our days on the trail unfolded, we read about these men in the trail registries. Sometimes they wrote notes to us, but we didn't see them again.

While Roadrunner and Tarheel hiked away on that second morning, Hikin' Fool lingered a few minutes to write in his journal. We didn't realize he was writing about us, and we didn't yet realize we were an oddity on the trail. Most hikers are singles or doubles. Hiking families, attempting the long distance we anticipated, were rare.

"I keep thinking your last name is Robinson." Hikin' Fool paused, pencil in his hand. "You know, like Swiss Family Robinson?" We laughed at the absurdity of it.

Then he had a brainstorm. "Swiss Family Wisconsin! That's what I'm going to call you in my journal."

Even though we are not Swiss, we were a family, and we originated in Wisconsin. That trail name stuck. As the hike continued, we joked about individual names. When our calorie intake could not keep up with our energy output, our body fat melted away. Joel was dubbed Walking Stick by his older brothers for his thin but sturdy legs, Tom was called Maxi Pack for his over-loaded pack, and Ben was called Roll Bar for the several times he tripped and did a graceful roll over his pack. But those individual nicknames were just family jokes. We were a unit. We were Swiss Family Wisconsin to everyone on the trail.

But to God, we were more than a unit. Like the sparkling stars in the night sky, God knew each of us by name, even our nicknames.

Vignette Five: Where's a Table?
Friday, June 8, Day Two

Give thanks to the Lord, for he is good, for his steadfast love endures forever

(Psalm 136:1 ESV).

At home, I had meticulously planned every meal, snack, and beverage. I'd dehydrated many of our foods, including fruit, beans, and even energy bars. Our meals and snacks were nutritious. I'd repacked packages into baggies to decrease weight and separated each meal into a bag, then packed the individual components of each meal into another bag for the day. Everything was labeled. Okay, I over-did the organization of it all. On the trail, reality hit, starting with supper the night before.

But now it was a new day and a new meal. I even had a new kitchen. With the other hikers gone, we moved into the shelter. The rough wood floor was much cleaner than the dust out-side, so we repacked and relaxed against walls. I sat on the step with the stove on the ground in front of me. My back was happy, and the stove was stable. According to the bag I'd labeled for June 8, we were going to have buckwheat pancakes.

Cooking breakfast required a mixing bowl and spoon, the skillet, cooking oil, and water. The healthy, delicious dry ingredients, like buckwheat flour, whole wheat flour, and powdered milk, were already in the breakfast bag. We also had little packets of McDonald's syrup that I had purchased in bulk and packed with the pancake bags.

Even without a table, we prayed our usual table prayer.

Come, Lord Jesus,
Be our guest,
And let this food
To us be blessed.
Amen.

But instead of eating together, only one person at a time got to eat, sitting on a step and balancing his plate on his knees. For two hours I fried pancakes on the little stove, one at a time. Those pancakes were filling, nutritious, delicious, and a blessing. For this meal, at least, I was a hero. The table God prepared was good.

Vignette Six: Traveling Heavy
Friday, June 8, Day Two

My guilt has gone over my head. Like a heavy burden, it is too heavy for me

(Psalm 38:4 EHV).

As I put away our array of cooking and eating utensils, a little niggle of doubt invaded my hero-mom facade. I pondered the conversation with Tarheel from the evening before.

His admonition not to carry any extra weight echoed in my head as we finalized our packing the next morning.

"Tom, I think we should turn around and hike the three and a half miles back to the road. We are carrying way too much stuff. We'll call our friend to bring the car. We'll unload the extra stuff and start over."

"You're kidding, right?" Tom's mouth hung open. "We've already wasted two hours of hiking time frying pancakes. We can't go back. We'll just ship the extra stuff home from the post office in Port Clinton where we'll be picking up our food shipment. It's only fifty-eight miles. How bad can that be?"

My kitchen wasn't the only extravagance. Tom's pack was borrowed, big, and old. From behind, he looked like a walking backpack with only his legs protruding. Tom had succumbed to the temptation to fill the giant pack, giving him a total weight of over sixty pounds to carry. In addition to his necessities, he had packed a heavy, metal Minolta camera, the kind that used film. He had twenty rolls of film, plus a telephoto lens and a wide-angle lens the size professionals used. Heavy binoculars, three books for casual reading, and a harmonica for campfire singing were stuffed into pack pockets.

"I carried a heavy pack in Vietnam," he had said at home. "This shouldn't be any different."

"No," I had argued, "Except it is twenty-four years later."

"Don't worry. We'll start slowly, and I'll get stronger as we go."

Ben, Caleb, and I carried between thirty-five and forty pounds each, depending on the food and water in our packs. On day two, we still had many pounds of food, enough to take us five or six more days to Port Clinton, Pennsylvania. Joel and Carla carried less. Joel had a few more supplies in his pack than Carla, but basically they carried their own needs plus the food of the day. Still, it was heavy for their small frames.

We had purchased light-weight sleeping bags for all of us, but they weren't the sleek, shrink-to-nothing sleeping bags of today. They didn't even fit into our packs. Instead they were rolled, tied, and stuffed into garbage bags to stay dry, then tied onto each pack, either towering on the top or dangling and butt-banging from the bottom.

We were traveling big and weighted down, each step an effort. Ten grueling miles loomed ahead. Because we were carrying too much, we were about to experience the spiritual portions of Psalm 38 in a truly physical fashion.

Vignette Seven: Fashion Statements
Friday, June 8, Day Two

Investigate me, God, and know my heart

(Psalm 139:23 EHV).

Fashion statements we were not. At 9 a.m., we tripped out of camp, late for seasoned hikers, but we weren't seasoned hikers, and we didn't have the look either. Some of our packs, our little stove, our water purifying pump, and our tents were new, but, for the most part, we were a motley crew with mismatched equipment, attire, and knowhow.

Gortex hats, breathable shirts, and wicking socks for six were not in our budget. As we stepped off on day two, wearing the same outfits as day one, I smiled at the impression we must have made. Nobody would mistake us for divas of outdoor design.

Tom and I wore our floppy Fleet Farm fishing hats with cotton handkerchiefs draped down our necks. That was supposed to discourage bugs. A blue Hard Rock Café shirt, blue-and-white seersucker pants from Goodwill, and black hiking boots completed my wardrobe. Tom in his khaki shorts and a Walleye Weekend T-shirt looked a bit more the hiker.

Ben and Caleb wore baseball caps. Joel tied a kerchief around his head, hiker-style, and Carla wore a self-decorated neon hat. High-top tennis shoes and a variety of T-shirts and pants completed their attire.

By noon, our clothes were drenched with sweat, and we were exhausted. We sprawled on the trail near a spring for our lunch break, devouring beef jerky, fruit roll-ups, and left-over pumpkin bread.

While the kids and Tom went to pump our canteens full of water again, I analyzed the handkerchief hanging from my hat down my neck. I understood its purpose, but sweat was trickling down my hot, veiled neck. I scanned the area for bugs. Compared to Wisconsin, the mosquitoes and flies were sparse. Perhaps the hanky would discourage ticks. Another drop of gritty sweat trickled down my neck. Who cared about tick protection? I shoved the handkerchief into my pack. It was an improvement in my style, but it didn't really matter to me. In fact, none of us cared how we looked.

Vignette Eight: Still Traveling Heavy
Friday, June 8, Day Two

I am drooping. I am completely bent over. All day long I go around mourning. Even my back burns with pain

(Psalm 38:6, 7 EHV).

The first obstacle as we descended Peter's Mountain was Shikellimy Rocks, a cliff of large, jumbled boulders to cross. A misstep could have caused a bad tumble.

"Watch this loose rock!" I warned. "Look out for that root."

I glanced at the kids' faces. "Quit rolling your eyes!"

I made a meaningless mom promise. "Okay, I'll lay off the cautions." Can a mom ever do that?

The downward trail steepened. Tom's feet slipped, and his sixty-pound backpack sat him down roughly on the trail.

We all gasped at what could have happened, what almost happened. Tom rolled over and pushed himself up. "I'm fine."

But none of us were totally fine. Because of our heavy packs and the steep trail careening downward, every step was pounding our toes into the front of our boots.

"My toes hurt," Carla whined.

"Mine too!" said Joel.

"I'm going to call this trail a toe stuffer," Tom announced. It was our first of many.

Shortly after a lunch of nuts and granola bars, we crossed Pennsylvania 325 and then Clark's Creek. It gurgled and sparkled below, beckoning us to its cool water.

"Let's go for a swim!" Carla eyes sparkled.

"No, sorry," Tom said. "We have a long way to hike yet today, so we better keep going."

The hiking mentality was already encroaching. A little voice in our heads whispered, "Keep going, keep going, keep going up one more hill, around one more turn, taking one more step. Let's get to camp where we can drop these heavy packs."

Descending Peter's Mountain was difficult, but climbing Stony Mountain, the next obstacle, was agony.

Vignette Nine: Uphill
Friday, June 8, Day Two

Yes, he will give a command to his angels concerning you, to guard you in all your ways. They will lift you up in their hands, so that you will not strike your foot against a stone

(Psalm 91:11, 12 EHV).

The guidebook promised that our ascent of Stony Mountain was on an old fire tower road. We expected an afternoon stroll up a smooth dirt road. This road, however, was rutted and had large boulders, strewn haphazardly about. We planted every step to avoid twisting an ankle or crashing over a rock. The guidebook offered no landmarks for the three miles to the summit, the longest three miles of my life—at least, until that day.

The boys pulled ahead of us, weary of waiting, and were out of sight before we thought to establish a meeting point.

Tom was having trouble with stamina too. He had none. He could only walk for fifty feet before stopping for a breather. His theory of getting in shape on the trail was dashed to pieces. On this day, he began using his infamous statement, "You go on ahead. I'll catch up."

Carla and I plodded steadily upward. The day was hot and humid. Sweat dripped down our faces.

"How much further to the top?" Carla's little face was flushed.

"I can't tell." I glanced at the trail, then I studied her for signs of overheating.

A few minutes later, she collapsed in tears. "I have to stop."

Equipped with know-how from our first aid course, I bolted into action, nudging Carla toward a patch of shade. "Sit here."

She slumped down. I fanned her face, sponged her with my wet handkerchief (an alternative use for the useless cloth), and gave her slow sips of warm water. Since we only had warm water, that didn't take much thought.

After a half hour, she felt better, and we started out again with me carrying her pack too. We were still alone.

As Carla and I struggled in this nonstop, upward trek, Tom caught up. Worried about the other half of our family, we shouted for them. After awhile, we heard Ben's welcome voice. He was sitting alone on a rock.

"Where are your brothers?" I asked.

"They kept going." He shrugged.

More worried than ever, we continued our slow, sporadic ascent, calling for Caleb and Joel, praying that God would keep them safe. Darkness crept closer as the summit of Stony Mountain loomed ahead. In the distance, a rumble of thunder rolled between the mountains.

Fear swept over me. "What if they hiked farther? How will we connect?"

We yelled their names louder. Finally, we heard an answer. At the summit, Caleb and Joel were perched on rocks in an open space between the trees.

"Hey guys, this looks like a good place to camp." Joel, unmindful of our concern, swept his hand around. "There are nice grassy, rock-free places for our tents."

Relief flooded through my body. Angels' wings swished in the trees overhead, or was that the wind picking up? Maybe both!

Vignette 10: First Storm
Friday, June 8, Day Two

They...were at their wits' end. Then they cried to the Lord in their trouble, and he delivered them from their distress. He made the storm be still. Then they were glad that the waters were quiet

(Psalm 107:27-30 ESV).

Our relief at being reunited was short-lived. The thunder rumbled louder. The refreshing breeze flashed into a roaring wind, and a drenching rain slammed down on us as the darkness of the evening descended.

"Get the tents up!" Tom yelled.

We bolted into action. The boys worked together on theirs, Tom and Carla on ours. The tents went up in record time. There was no time to set up a third tent for our packs, so I set them in a circle, like a wagon train waiting for an attack, and threw a piece of plastic over them. Hoping to keep their precious contents dry, I tucked the plastic under each pack.

But quick action did not keep us from getting drenched. Soaked to the skin, breathless, we dove into our tents.

"Good teamwork," Tom yelled to the boys through our tent walls.

"Wow! That was awesome!" Caleb's voice drifted back.

"Thanks, Carla and Mom." Tom nodded at us and slumped onto the hard tent floor.

"I'm a little chilly." Carla shivered and rolled into a fetal position, wrapping her arms around her legs.

"I just want to flatten out my back." I collapsed, coaxing out the strain of carrying a pack for ten miles. "This feels so good."

But it didn't feel good for long. Relaxing was our only comfort. We hadn't had time to grab food or dry clothes or even our sleeping bags before the storm hit. So we shivered in our stark, flimsy tents with no extra comforts while the storm blasted and our tents' walls whipped back and forth.

Lightning flashed. Thunder crashed on its heels.

"This is probably the worst place to be camping." Tom's voice quavered as he shivered. "On the top of a mountain, under trees, huddling in wet nylon with aluminum tent poles. We are an easy target of the next flash."

"It's where we all came together. We didn't really have a choice." But I knew the statistics. More people die of lightning strikes than of being hit by a tornado.

I grabbed Tom and Carla's hands, and we clung together. "Dear Lord, keep us safe."

The storm raged for one and a half hours, and then it rumbled off.

Vignette 11: Storm Within
Friday, June 8, Day Two

Do not forsake me, O Lord. My God, do not be far from me. Hurry to help me, O Lord, my salvation

(Psalm 38:21-22 EHV).

When the rain stopped, we crawled from our saggy tents into the soggy world, assessing the damage. Wind-swept leaves and twigs cluttered the ground. I studied the circle of plastic-covered packs, relieved to see my hasty protection had held.

"It looks good! Still, I'm afraid to uncover them."

Wet packs would mean wet sleeping bags, spoiled food, and damp clothes. I tugged at the plastic. Ben grabbed the other side of the plastic, and we gently rolled it back.

"Hallelujah!" I danced in a circle. "Everything is dry."

We giggled as we shivered and shed our soggy clothes and luxuriated in the warmth of our dry shirts and pants. But Tom could not stop shivering. He looked purple from his ears to his hands.

"I'll be fine," Tom said. But the rest of us were concerned about hypothermia. The storm had dropped the air temperature to the 50s. How could we go from heat exhaustion to hypothermia in a few hours? But that was a reality of the trail.

"Remember that wood pile back on the trail?" Ben turned to his brothers. "It was covered with plastic. Let's go get some."

The three boys tore off, back down the trail, and in a few minutes returned with huge armloads of dry wood. After many futile attempts and many worried glances at Dad, working together, they got a fire roaring.

As Tom sat by the fire, warming himself, his color improved. The mental state of all of us improved.

I cooked beef jerky in rice over the open fire. We ate it from the kettle because, according to the guidebook, there were no springs for another eight miles, and we didn't have any water to spare for washing dishes. Granola bars and hot chocolate topped off our meal. After supper, digging in my pack, I found an old peanut butter and jelly sandwich I'd thrown in the day before. It had been left from our car lunch three days earlier. At that time, when I had offered the sandwich to a "starving" kid, he was more interested in stopping at a DQ.

Today I waved the sandwich in the air and asked, "Who wants this old thing?" Four ravenous children pounced on me.

With the fire still roaring, I was thrilled to burn two days of accumulated papers and packages. Since there are few trash cans along the AT, hikers must pack out all garbage. Life was looking better. My pack was cleaned out, we were well fed, and I was feeling a little smug, thinking that we had handled that crisis. Just then, a distant rumble echoed through the hills. Instantly, I was reminded that we hadn't handled anything well. We were in God's hands.

~Distance hiked on day two: 9.4 miles

~Peter's Mountain Shelter to the summit of Stony Mountain

Vignette 12: Grand Finale Storm
Friday-Saturday, June 8-9, Days Two-Three

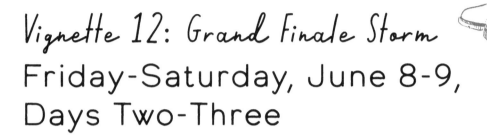

> The God of glory thunders. The voice of the Lord breaks the cedars. The voice of the Lord flashes forth flames of fire. The voice of the Lord… strips the forests bare and in his temple all cry, 'Glory'
> *(Psalm 29:3, 5, 7, 9 ESV).*

Another rumble sounded in the distance, then another. Closer.

Joel, always thinking, said, "Since we don't have much water, let's set out kettles to collect rainwater." Joel and I scurried to do that.

Tom and Caleb took care of the food, placing our individual food bags into a garbage bag and then into a nylon sack which held Tom's sleeping bag during the day. Hmm, how smart was that, adding the scent of food to Tom's sleeping bag? Such novices! After all the food was in the bag, Tom tied it shut and threw the rope over a branch. Together, Ben, Caleb, and Tom pulled the bag into the air, high enough so that bears could not reach it from the ground and far enough from a tree trunk so bears could not climb up and snag it. This became a nightly routine.

Our food bag was also mouse-proof, thanks to Tarheel. The previous night, Tarheel had pulled the lid of a tobacco can through the tightening cord of our food bag and pushed it down to where the bag tied shut. "When mice climb down the cord they'll be blocked from getting into that little hole and into your food bag. Little critters can do as much damage as big ones." Protected from bears and mice, our food was secure.

Ben and Carla put our packs into a secure circle and covered them with plastic again. At that point, we decided that a third tent was unnecessary and we'd send it home when we got to Port Clinton. Traveling light was becoming a focus.

As the first sprinkles fell, we felt prepared and secure. We zipped ourselves into our tents, warm, dry and content. Then rain fell. And fell and fell. And thus day two of our hike came to a close. How could so much happen in less than two days of hiking? I was thankful that our family was working together in crises, and I praised God for keeping us safe through many dangers. Then my eyes closed in contented sleep.

At 1:30 a.m., we were jarred awake when a blast of thunder rocked our tents. The storm was right on top of us, right on top of Stony Mountain. Even though we were better prepared than earlier to ride out this storm, it was still terrifying. We were still in wet, flimsy tents with aluminum poles. We were still on high ground with trees above us to attract the lightning. As the storm intensified, lightning flashed, followed immediately with thunderous crashes. Until 3:30 a.m., we cowered in our tents and prayed. Only God could keep us safe in this terror. He did.

For days, and even weeks later, the big storm was the hot topic of conversation with other hikers. We compared stories about where we were and how we had fared. The other hikers did not envy our position hunkered down on the summit of a mountain, but we felt blessed by it.

Vignette 13: Running Behind
Saturday, June 9, Day Three

Let me hear about your mercy in the morning, for I trust in you.
Teach me the way that I should go, for I lift up my soul to you

(Psalm 143:8 EHV).

We awoke at 6 a.m. and strung lines to dry clothes and sleeping bags. With relief, we discovered we were not that wet.

Joel helped me filter the rainwater for cooking oatmeal, and once again we gathered around with our spoons and ate from the kettle. Water couldn't be wasted on washing dishes. Eight miles of hiking down the trail was a long way to go for our first water spring.

I also served power-packed, perfect-protein oatmeal bars that I had made and dehydrated at home. This was actually a successful attempt at preparedness. Did that make two out of twenty attempts?

With memories of yesterday's grueling climb still fresh in our minds, we were jolted to reality by two joggers who effortlessly breezed through our camp. They had run up the old fire tower road, the horrible trail that had almost defeated us.

"I can't believe you ran up this mountain," I shouted.

"This is just the start." One runner trotted backwards, not even out of breath, before he sprinted away.

"Are we in shape or what?" I asked nobody in particular. Tom shook his head and led the way out of camp.

A half mile down the trail we came to a little stream called Rattling Run. It wasn't mentioned in the guidebook as a water source, but its water was sparkling and clear. We pumped our canteens full. What a relief to have plenty of water again.

Since it was Saturday, we encountered weekend hikers. Three ladies passed us.

"We are planning to hike fifteen miles today," one called over her shoulder.

"Our reward goal is a little store about ten miles away that sells ice cream," another said as they sped off.

"Can that be our goal too?" Carla pleaded, so we adopted it.

"They weren't carrying heavy packs," Tom mused. "That's why they could zoom past us."

We all agreed.

Soon a troop of Boy Scouts hiked past us. They were carrying heavy packs.

"They're probably not as heavy as ours," Ben said.

We all agreed again.

The trail followed an old stagecoach road for about seven miles. We expected it to be an improvement over the fire tower road of yesterday. It wasn't. Only stagecoaches with balloon tires and four-wheel drive could have maneuvered that road. Huge boulders and various sizes of rocks were strewn helter-skelter. Pennsylvania's boot-killing reputation was rearing its ugly head. Sometimes the trail disappeared in the rock jumble.

"Is this the trail?" Tom turned in circles, lifting his hands in exasperation. "Where is a white blaze trail marker?" Just then, more hikers sprinted past.

Vignette 14: Running More Behind
Saturday, June 9, Day Three

He leads the humble in what is right, and teaches the humble his way

(Psalm 25:9 ESV).

All morning, I had been enthralled with the beautiful bushes lining the trail and parading up the hillsides. The bushes were six to ten feet tall and draped with clusters of delicate, umbrella-like blooms in shades of reds, pinks, or whites. So when the Boy Scouts stopped for lunch, I paused to ask an expert, a Scout, about the flowering bushes.

"They're mountain laurels, ma'am, the state flower of Pennsylvania," the Scout said.

"They're so elegant. How surprising that they grow wild here. Mountain laurels could grace the front lawns of Southern plantations."

"Yes, ma'am, but they also grow along the AT and on many hillsides in Pennsylvania."

The other Scouts nodded affirmation. What polite, smart boys.

While I diverted the Scouts' attention, the rest of my family slipped past.

"Maybe we can stay ahead of them." Caleb picked up his pace.

"Maybe the guidebook's description of uphill along a creek bed will be easier." I bolted off.

Foolishly, we expected that the trail would be beside the creek or that the creek bed would be dry. Wrong. At first we picked our way up the creek, jumping from rock to log to rock to avoid getting our feet wetter.

The Boy Scouts again passed us, sloshing through the creek. Caleb's shoulders slumped.

I mumbled as they splashed by me. "What would your mothers say about getting your shoes wet?"

But shortly, we followed their example. Our feet were already swimming in our boots from hiking through wet grass all morning. A new hiking theory was born. When one's feet are soaking wet, they can't get wetter. And sloshing up the creek bed was easier than balancing from rock to log.

By late morning, several more hikers had passed us, and then the ultimate humiliation descended on us. We stopped to sign a trail register and were pleased that Roadrunner and the F & S Boys had written personal greetings to us. But then Tom noticed the time of their entries.

"They signed in here yesterday before lunch. That means they hiked in a morning the distance we hiked in a whole day plus a morning."

Our slow pace embarrassed us and crushed the envy we were feeling. We needed to adjust our competitive spirits and return to reality. We were novices, we were overloaded, and we were a family of various abilities. God loves a humble spirit, and we needed to work on ours.

Vignette 15: Not-So-Nice Doggie
Saturday, June 9, Day Three

The Lord is the stronghold of my life; of whom shall I be afraid

(Psalm 27:1 ESV)?

Besides humble pie for lunch, we ate peanut butter and jelly crackers with dried apricots. We perched around a big rock on the side of the muddy trail, squatting or sitting on little rocks.

Joel balanced his whittled walking stick across his legs. I studied it. "I like how you scraped off the bark to make a clean, white surface, and your leather loop through the top makes it easier to carry."

We were admiring his handiwork when a shaggy man and his unleashed, sturdy brown dog, standing two feet tall and weighing half as much as me, sauntered in from the north.

"Hi folks!" the man said. "Ali and I are doing the AT from north to south."

Ali wasn't interested in introductions. The dog lunged toward Tom, cornering him against a soggy embankment, and licked Tom's hands and legs. Tom is infamous in our family circle for not liking dogs, probably because he was attacked many times during his childhood newspaper routes. As Tom gingerly patted the drooling beast, the kids and I exchanged fearful glances. Tom's eyes were big with terror.

I had an inkling of Ali's breed. "Is that a pit bull?" I inched back a bit.

"Ali sure is." The man patted Ali's big head and grinned. "The good kind of pit bull." Ali wiggled, snapped his head around, and glared at me. I stepped back again.

Good kind or not, I knew that the breed came from the United Kingdom where, historically, pit bulls competed in combat sports like dog fighting. The breed has changed in America to include a bit of terrier. Sometimes they are sweet pets, but they are responsible for more dog attacks in the U.S. than any other breed. Here's the clincher. They have the biting power of 2,500 pounds per square inch. Lions and tigers only bite at 1,000 pounds per square inch.

I planned to keep my distance, but Ali was diverted from me. He spotted an enticing toy, Joel's walking stick. Ali pranced over to Joel and nibbled on the bottom of it. Joel leaned back. Ali continued to snap his way up the stick. Then he clenched the stick in his teeth and shook his head.

"If the dog wants your stick, Joel, let him have it," I hissed.

Joel released his grip. The dog pulled the stick free, chomped down on it and chewed it into splinters. We gasped in one united breath.

"Well, I best get going." The man ignored the drama and took off south, whistling for his dog to follow. Ali bolted behind, scattering splinters in his wake.

"What just happened?" Ben glanced with down-turned mouth at his little brother.

Joel shrugged and tapped his head. "I think I just discovered a money-making idea. I'm going to make walking-stick toys for pit bulls. They seem especially attracted to hand-whittled ones."

A brief moment of stunned silence followed, and then we all burst into hysterical giggles.

We never saw Ali and his owner again, but we did hear tales of a pit bull running wild near the Appalachian Trail.

Vignette 16: Guidebooks
Saturday, June 9, Day Three

Make me to know your ways, O Lord; teach me your paths. Lead me in your truth and teach me, for you are the God of my salvation; for you I wait all the day long

(Psalm 25:4-5 ESV).

While we struggled to accept our slow pace, we also struggled with trail descriptions. This was before cell phones and GPS. Our only options were paper maps and guidebooks. Sadly, we were confused by the *Appalachian Trail Guide to Pennsylvania.* We thought we had the best, including guidebooks for each state, official trail maps, the *Appalachian Trail Data Book 1990,* and the *1990 Philosopher's Guide.* We even carried road maps for the bigger picture.

Why were we confused? Part of the reason was that there were ten trail guides for the entire AT. The AT in Pennsylvania has its own book, but it was subdivided into fourteen sections. Each section in the guide could be followed from north to south or south to north.

"Wait." I squinted at the guidebook, puzzled more than once. "I was looking at the right section but the wrong direction."

The guidebooks have been written and rewritten by an army of volunteers. They have made corrections and adjustments as the trail has evolved since its completion in 1937. Various writing styles abound.

These same groups also maintain the entire trail. Writing guidebooks and maintaining the trail have been formidable undertakings. The guidebooks are extremely detailed, so much so that I often carried one in my hand for easy reference.

This morning at the summit of Stony Mountain, we started at 3.3 miles into Section 7, heading north. At 3.9 miles we crossed Rattling Run, the unexpected water source.

The guide continued confusing us, mentioning an old stagecoach road at four miles, which was unrecognizable as a road, and noting the former Yellow Springs Station at six miles, which was nonexistent. It also failed to mention that the trail was in the creek bed not along it.

With the landmarks so difficult to recognize, we'd whoop with delight when a white blaze appeared, confirming that we were still on the trail.

Later in the day, we recognized Rausch Creek, as it tumbled and foamed like a first-rate mountain stream. Shortly, the blue blazes, plus the guidebook, led us to the three-sided Rausch Gap shelter. Known as the Hilton of the AT with its rock and big-timber construction, it sports two levels, one for sitting and one for sleeping. A young man was resting there, hoping his swollen leg would allow him to hike again the following day.

"This is a good place to stay." I glanced at Tom and pointed toward the nearby spring and the grassy areas for our tents. "There's even a privy. We've already hiked eight miles today. Let's call it a day!"

Tom countered. "I see in the guidebook that there's a spring in two-and-a-half miles. We still have some good daylight. Let's head to the next spring."

We never found the next spring. Perhaps we missed the sign for it. Where was reliable information?

Vignette 17: Calorie Problems
Saturday, June 9, Day Three

We went through fire and through water; yet you brought us out to a place of abundance

(Psalm 66:12 ESV).

Our pit-bull-hosted lunch, hours before, had been simple and nutritious but not enough. By the time we'd hiked to that invisible spring in the prior vignette, we'd put in ten grueling miles. We were tired and hungry.

"Hey, according to the guidebook, we'll be at the little grocery store in a mile," Tom exclaimed. "Remember? The three ladies told us about it this morning. That was one of our goals. We can do that. Let's go! Ice cream ahead!"

"Yea, let's go." Ben cheered along.

Tom. Another spring only two miles away. Groaning with protruding tummies, we hoped we'd reach our next destination before the sun set.

We found the spring and a place to camp beside Swatara Creek, but we did not feel well. Oh, those rich, ice cream calories were heavy. Ben and Caleb moaned with headaches. The rest of us felt bloated but forced down Ramen noodle soup for dinner before bed. As darkness settled in, a commotion erupted in the boys' tent, followed by a zip-zip of the door.

"Hurry! Get out of here!" Ben yelled, followed by a suspicious coughing outside.

"Everybody okay out there?" I asked through the tent wall.

"Yea," sputtered Caleb. "I was sick, but I'm OK now."

And so we learned that needing more calories isn't remedied by eating lots of ice cream, a lesson we'd relearn multiple times on the hike.

~Distance hiked on day three: 13.6 miles
~Stony Mountain Summit to Swatara Creek

Forgetting about the rice pudding that was thickening, heart pounding, I tore up the trail, meeting Carla alone. "What's wrong," I gasped. "Where's Joel?"

"Oh, he's at the bridge. We were wondering what was taking you so long."

Miscommunication. They thought we'd be joining them. We thought they'd come back, and we'd pick up our dry clothes en route.

Back at camp, Ben had smelled the rice pudding burning and had turned it off. He saved the meal. Almost. It only tasted a little scorched once we smothered it with raisins and cinnamon. When it was blended together, we hardly noticed the burned specks. There was plenty of nutritious rice pudding to energize us, and it did.

Caleb had successfully caught three little sun fish. He cleaned them. I fried them. We each got two bites of fresh protein to complete our brunch.

By the time we packed up, our relaxing morning was not fun or restful. The pretty sun had turned hot, and we were all frustrated with our pokiness. The Boy Scouts passed us, the invalid hiker passed us, and it was already afternoon. Our planned rest was not heavenly, and we were not feeling like saints.

Vignette 19: Minor Troubles
Sunday, June 10, Day Four

[The Lord] is their stronghold in times of stress

(Psalm 37:39 EHV).

The Waterville Bridge, across Swatara Creek, is a historic iron bridge, now only used by the AT. Carla and Joel had spread our hand-scrubbed-but-not-really-clean clothes there to dry. While we paused to pack up, Tom added more mole skin bandages to his feet. Blisters were popping up in multiple places on his feet and on ours as well. Yesterday's long, foot-pounding hike roused blisters that would plague us for weeks.

Underway again, we grinned at each other as we hiked past the Boy Scout troop eating lunch beside the creek. Seeing them spurred us on. We then crossed under Interstate 81 and climbed up another mountain. It was a lung-expanding, one-mile climb, but we were hoping for a breather and a spectacular view at the top. We got the breather, but the crest of the mountain was tree-covered and overgrown with vines.

Tom scowled through the jungle. "Can you believe, after all our efforts, there is no view at the top?"

"I'm more concerned with the view ahead." I nodded toward the trail. The vines had woven themselves into a mesh with clumps of three leaves protruding.

Joel raised an eyebrow. "You know what they say about leaves of three."

"Let them be!" Carla shouted.

"Well, it could be poison ivy." Tom studied the vines. "But there's no way around. Let's push our way through and pray that nobody reacts." Dealing with an itchy, oozing rash on the trail could be a big problem. Thankfully, poison ivy didn't trouble us later.

Fallen trees, sprawled across the trail, and treacherous rocks replaced the poison ivy as problems. One time we lost the trail completely and scanned the area for the trail-marking white blazes on trees. The next blaze was painted on rocks.

"Of course it's painted on rocks here," Tom said. "These rocks are so camouflaged that we have to keep our eyes on the trail constantly to know where we can safely put a foot for the next step." Rocks on the AT were recurring problems.

We lunched on a nut mix with M&Ms plus dried fruit. The Boy Scouts passed us. Later, as we hiked along the crest of the mountain, we passed the Boy Scouts setting up camp. I looked longingly at their site but knew we couldn't stop here. There was no water available, and we'd be in trouble, camping without water.

Little problems were building up. We were hot and tired, frustrated, despising this rocky, tree-strewn trail, trusting our own abilities.

Vignette 20: Real Trouble
Sunday, June 10, Day Four

But [God] you do see. You notice trouble and grief. You take it into your own hands

(Psalm 10:14 EHV).

When real trouble jumps up, all little troubles vaporize. It was still Sunday, and we were still miserable. We were beginning our descent off the mountain crest. The overhead trees had disappeared, and the hot sun was beating down when fourteen-year-old Caleb took the lead. He was absorbed in swinging a stick at flying insects. Suddenly, he stopped swinging, took a panicky leap ahead, and screamed, "Snake!"

Caleb had heard the rattle, but with his downhill momentum, his only option was to leap over the coiled rattlesnake. The rest of us skidded to a halt. We stood there with hearts pounding and knees quivering. In the middle of the narrow trail a rattler separated Caleb from us. This was a big problem, but not as big as it could have been or could get.

"I don't have a snake bite kit." I grabbed Tom's sleeve.

"Caleb is fine," Tom said. "We're all fine. Let's just calm down and think."

"Remember what the F & S Boys told us?" I asked. "Let's pound a stick on the trail, and the snake will slither off."

We beat the ground with sticks. The snake did not budge. Tom grabbed a bigger stick and pounded it closer. Still, there was no reaction from the snake. Ben threw stones. The snake held its ground.

"Why do we have a snake that doesn't follow the rules of nature?" I moaned.

Several minutes passed, and all our efforts to coax that snake along failed. Maybe it was as nervous as we were. While Tom and I consulted with each other, sixteen-year-old Ben, impatient with our futile attempts, plowed through the underbrush on the side of the trail to get to his brother.

"I hope that snake is not a social creature with friends and family nearby," Tom muttered as the rest of us followed Ben.

When we were well past the snake, Tom threw a rock back. The snake darted off, looking as relieved as us to escape.

With the danger averted, we breathed a prayer of thanks and hiked on. Already, it seemed like just another day on the AT. But our radar did perk up for future encounters with snakes. And there were more ahead.

Vignette 21: Darkness and Rocks
Sunday, June 10, Day Four

For who is God besides the Lord? And who is the Rock except our God? This God wraps me with strength and makes my way perfect. By making my feet like those of a deer he enables me to stand on high places

(Psalm 18:31-33 EHV).

When we reached the side trail for Blue Mountain Spring 200 yards below us, we had already hiked seven-and-a-half miles that Sunday afternoon. The boys went down to the spring to fill our canteens while Carla, Tom, and I waited. We were beginning to worry about how long it was taking when the boys emerged from the side trail.

"We were helping that invalid hiker from Rausch Gap start a fire," Joel explained. "His knee is swollen again, and he's having trouble walking."

"This could be a good place to camp." I glanced at Tom. "We'd have plenty of water, and we'd be available to help the hiker."

"But it would be nice to get to the hostel only four miles ahead."

"It is 6 p.m. already and will soon be getting dark."

Ben intervened. "Let's toss a coin." The coin said "stay."

But we decided we didn't need to rely on the chance of a coin toss. We did want to go, to get to the limited comfort of a hostel, so after a quick snack, we took off again.

The AT was easy for the first two miles. We were jubilant until the trail became rockier with difficult climbs.

About this time, we met a young Mennonite couple strolling on the trail. The girl had on a calf-length blue dress and a white prayer cap. The guy was neatly dressed in a shirt and trousers. I'd grown up around Mennonites and knew they were conservative Christians, sometimes confused with Amish. Mennonites, however, do not live as simply as Amish. They even drive cars.

These young people smelled delightful, like they had just stepped out of an Ivory soap bath. I wondered what we stunk like with only a scanty sponge-off in the past four days. They smiled, said hello, and continued walking the opposite way we were going.

As we pushed on, we were treated to an occasional view below, and we allowed ourselves quick stops. Darkness was threatening when the trail became full of the remnants of massive boulder slides. Carla, Joel, Caleb, and I pulled ahead, anxious to get to stable footing while we could still see. In the looming darkness, we were grateful the white blazes were again painted on the rocks.

At eleven miles for the day, we paused at Kimmel Lookout, named for Dick Kimmel, an AT trail worker for more than forty years. Here on a grassy hillside, two more Mennonite couples sat on blankets, feasting their eyes on the farmland far below. We exchanged pleasantries and plodded on.

The contrast between these serene couples and our family's mounting feelings of desperation was puzzling. But then, they were on familiar terrain, had cars awaiting them in the parking lot and cozy homes to return to. We didn't know what was ahead.

Before we got to Pennsylvania Route 501, we tripped over one more slippery, rockslide and finally climbed the wall into a parking lot. The solid rock of the parking lot was a comfort. Ahh! Safety!

Vignette 22: Friends and Fiends
Sunday, June 10, Day Four

Keep me safe, Lord, from the evil man. Protect me from the violent man, who plans evil in his heart

(Psalm 140:1-2 EHV).

Caleb, Joel, Carla, and I sat on the rock wall of the parking lot, waiting for Tom and Ben to arrive. Only the soft glow of an-already-set sun gave us dusky light.

"I wonder how far back they are," Joel mused.

"I hope they can see well enough to climb over that last rock scramble." Caleb frowned.

"Do you think we should go back and look?" Joel stood, hands on hips.

"I think we should." Caleb stood up too.

To prevent any foolish rescue attempt, I decided to send Caleb and Joel ahead, across the highway and only a short walk farther to the hostel. "Why don't you two just follow the trail there? You can get settled in."

"That's a great idea, Mom," Caleb said, and off they went.

Now it was just Carla and me sitting on the rock wall, waiting, alone. I began to question my decision to send the boys on ahead, alone. They had become quite independent and reliable on the trail, so I didn't worry about the idea until it was too late to reverse it. Now Ben and Tom were alone, Caleb and Joel were alone, and Carla and I were alone.

Before total panic took over, the three Mennonite couples returned to the parking lot. "Did you see the rest of our family?" I asked.

One of the young men answered, "No, we didn't see them, but we did hear them only a short distance behind us."

With a sigh of relief, Carla and I continued our wait. The Mennonite couples strolled to their cars, but oddly, they did not leave. Just then, several undesirable-looking characters rumbled into the lot in their rusted-out Ford truck. They rolled by, ogling us. One glanced at the couples opposite us, watching from their cars. With stones flying, the truck squealed its tires out of the parking lot.

Shortly, Tom and Ben appeared. As they climbed over the rock wall, the couples smiled, waved, and took off. Were they watching out for us, making sure we were safe before they left? Maybe they were guardian angels in Mennonite clothes.

Vignette 23: Blessed and Secure
Sunday, June 10, Day Four

I have set the Lord always before me; because he is at my right hand, I shall not be shaken. Therefore my heart is glad, and my whole being rejoices; my flesh also dwells secure

(Psalm 16:8-9 ESV).

Shelters along the Appalachian Trail were varied. Some were three-sided. Some were made from big timber, and some were dilapidated. Most did not have doors. Hikers cannot reserve shelters. All are on a first-come-first-served basis. But our first Sunday on the trail, our supposed day of rest was coming to fruition. The shelter was available.

The hostel was only a shed with six wooden bunks and a plank table with candles and wax drippings dotting its top. There was a door but no electricity. A privy and a solar shower were outside. Nearby, a hose strung from the caretaker's house provided our water. This hostel wasn't even a one-star. It wasn't home either, but it was fabulous.

First of all, we didn't need to set up our tents, a time and energy saver. We also had a real roof over our heads, and the ceiling was high enough to stand up. We had running water that

we didn't need to filter. We could cook at a table and eat by candlelight. We could throw garbage away and not pack it out. We didn't need to find a spot behind a tree or rock to go to the bathroom, and our privy even had toilet paper.

Caleb and Joel had already claimed two top bunks with their sleeping bags. Ben jumped onto the last one. Carla and I didn't care which wooden bunk we had. A hard wooden plank is a hard wooden plank. But we did care about a shower and called dibs on the solar shower first.

Now, one would expect that a solar shower, with a water tank heated by the sun, would be a warm shower. Wrong. It was freezing. We got wet. Screamed. Stepped out. Soaped up. Rinsed off. Screamed. Dried off. Put on clean clothes and felt wonderful.

"Do we smell like we just had an Ivory soap bath?" I waltzed into the hostel.

Tom sniffed. "Almost, except for the slight scent of your river-washed clothes."

While the guys cleaned up, I cooked stove-top stuffing and then added a can of Spam cut into chunks.

"This is a great meal, Mom," Caleb said. While I basked in the compliment, he added, "Did you know that the caretaker has Klondike bars and Cokes for sale? Do you think we could buy some?"

"Of course."

One more comfort of this hostel. The kids relished every swallow of their treats, climbed into their bunks, and were sleeping in seconds.

That ended our first Sunday on the AT, our day of rest that went awry and then once again became restful. A shed with six wooden bunks became a mansion. A frigid shower washed us clean. Spam and Klondike bars were foods for feasting. Feeling blessed and secure, we fell asleep.

~Distance hiked on day four: 11.3 miles
~Swatara Creek to Pennsylvania 501

Vignette 24 Worms and Webs
Monday, June 11, Day 5

He spoke, and the locusts came, and grasshoppers without number.
They ate every green plant in their land

(Psalm 105:34-35 EHV).

"I declare this a buckwheat pancake morning," I announced, even though it wasn't on my menu for the day. We had a table in our hostel, which meant I could stand by the stove for hours and cook. I had packed this lengthy-prep breakfast, and we were going to eat it, no matter how long it took. We topped off our nutritious meal with more nutrition, a cup of Gookinaid, a powdered health drink mixed with hot water. It tasted as bad as its name implied, but we were doubly fortified for the day.

We had only hiked a short, fortified distance when we noticed gypsy moth caterpillars everywhere. They were in a feeding frenzy in the trees overhead. We learned that the larvae go through five or six stages, each time emerging with a bigger appetite. Usually they'll only eat at night, but if there is a high density of them, they'll eat all day too. Obviously, we were in a high-density area.

It was eerie because there were few leaves left on the trees, and the air smelled like autumn from the decaying leaves carpeting our trail. Three years of defoliation will kill a forest.

The first stage larvae were hanging by silken threads, moving with the wind. These silken threads were why the gypsy moth was introduced in Boston in 1869. It was an experiment in making silk. The experiment was a horrible failure. Now the population of gypsy moths has exploded as they slowly chomp their way across the country.

Ben was the lead person in our hiking line-up when we encountered these young caterpillars swinging on silken threads. And he got to experience these silken threads firsthand, or should I say, face first, as webs draped over his head. After he had been in that position for awhile, Ben sputtered, "Somebody else take the lead now."

We all did reluctantly take our turn. There was an advantage to being in front. That person could bat the swinging larvae out of the way, an easy form of entertainment as we plodded along.

"Hey, watch this," Caleb shouted from his lead position. He carefully batted the swinging caterpillar so that it swung back and hit Carla.

"Stop it," she yelled but laughed. We all did, even though it was disgusting.

Those caterpillars managed to land on our backpacks too. Most of the day, one or two would be climbing on a pack. We took turns snapping them off each other. Sometimes they managed to wiggle under our pack straps, and we'd find them later, smashed into our shirts, leaving permanent gut stains.

I grabbed a tree for balance and smashed a hatch of emerging larvae. "Yuck." With a shudder, I wiped my gooey hand on the grass.

Even when we paused to eat our granola bar and peanut butter lunch, we couldn't escape the caterpillars. They dropped on our backs. They crawled up our legs. The kids snapped the caterpillars at each other.

"Enough of that!" Tom said. "Let's think of reasons we should be thankful for these hungry caterpillars while we thank God for our food." Thoughtful silence.

Vignette 25: Persistence at Shubert's Gap
Monday, June 11, Day 5

Indeed, I call to you because you will answer me, O God. Turn your ear toward me. Hear what I say

(Psalm 17:6 EHV).

Since the trail was still rocky, we moved slowly. In the early afternoon, we came to a park-like hillside in Shubert's Gap. Springs bubbled up, weaving into little brooks which tumbled to a deep pond below. Dark green moss covered the hillside in soft carpeting, and there were wooden platforms a few inches off the ground for tents.

"Tom, this is the place I've been dreaming about. It's so beautiful." I twirled around. "Let's set up camp."

"Are you kidding? It's too early to stop hiking." Tom shook his head. "We've hiked less than four miles today."

"Remember all those days you made us hike farther than was good for us?" I countered. "Maybe your blisters from those long days of hiking need a rest today."

"If we are going to hike 500 miles in two months, we can't be dillydallying."

"Didn't we also plan to enjoy our days on the trail, not make every day a forced march? I think staying here would refresh my soul." I stepped back, afraid I might have gotten too dramatic.

But Tom missed my hesitation and crossed his arms. "We'd be wasting six hours of daylight."

"You could catch up writing in your journal." Keeping current with his journaling was always his goal.

I could feel Tom's resolve weakening, so I pressed the matter. "The boys could go fishing in the pond. We could play cards. We could have a peaceful evening around a campfire, all things we hoped to do on this family adventure."

The kids carefully watched the discussion, turning their heads from one parent to the other like they were watching a ping pong match. They waited expectantly. It was Tom's serve. He sighed and gazed heavenward, knowing he'd lost the match, knowing that he wanted to give us this gift.

We all jumped up and down, cheering wildly. My persistence had paid off. I just needed to ask and then ask some more.

Vignette 26: Dreamland
Monday, June 11, Day Five

He makes me lie down in green pastures. He leads me beside still waters. He restores my soul

(Psalm 23:2-3 ESV).

"Okay, let's claim the platforms for our tents by setting them up, then we can relax and have fun the rest of the afternoon." Tom was embracing the idea of time off.

Quickly, our little tent-houses emerged. Having wooden platforms made the set up easy. There was no searching for a level surface, no brushing stones aside, and no dirt right outside the door.

"If it rains, we'll be above the muddy ground," Joel pointed out.

The only drawback was the caterpillars. Even before the guys had the tents up, they were covered with caterpillars. Our packs, set in their usual circle, were also soon covered with those squishy little creatures. But we began to appreciate them.

"They're not as bad as flies or mosquitoes," Tom said. We all agreed. Given a choice between flies or mosquitoes or caterpillars, we'd pick caterpillars. They didn't bite, buzz, or fly. We were content to put up with them.

During the idyllic afternoon, the boys went down to the pond to fish, wash up, throw rocks, and capture little creatures, while Tom settled in with his journal.

Carla and I heated water over the fire to wash our hair. We hadn't been brave enough in the cold solar shower the evening before. Now, she poured warm water over my head, and I poured warm water over hers. We glowed with contentment.

The Boy Scouts arrived in the camp above us. They were distantly friendly. Maybe they were surprised to see us in camp ahead of them. Maybe they felt a little happy to see us. During the past days, they had become to us like that neighbor down the street—not your best friend, but comfortably close.

Because we had a campfire, I again diverted from my planned menu in favor of the long-cooking Appalachian Trail Mix: brown rice, barley, and lentils spiced with chili seasoning. It simmered away for more than an hour and didn't use one whiff of fuel.

"Come and get it," I called.

"Smells good," Caleb said.

Ben swallowed a bite. "It tastes okay, but it's not my favorite." He shoveled in a second bite.

"Well, it's very good for you," I said. "It's healthy, filling, and full of protein. Pull up a rock to sit on and dig in."

Carla tried to snuggle up on my rock, but she didn't fit. "That's okay, Mom. I'll just pull up another rock."

It was sunset when we climbed into our tents and slipped into our sleeping bags with contented sighs. Then I heard the boys' tent flap zipping open.

"What's going on?" I asked.

"Just throwing out a blessed caterpillar," Caleb said, followed by the sweet sound of their laughter.

~Distance hiked on day five: 3.5 miles
~Pennsylvania Route 501 to Shubert's Gap

Vignette 27: Humbled by Oatmeal
Tuesday, June 12, Day Six

For you save humble people, but you bring low the eyes of the arrogant.
(Psalm 18:27 EHV).

We awoke refreshed, energized, and refueled. Our legs were feeling stronger. We were trail savvy and ready to roll. Even the caterpillar doo-doo, pelting us from the trees above, couldn't daunt us.

"Let's see if we can break camp before the Boy Scouts do," Caleb challenged us. With renewed vigor, the boys collapsed, folded, and rolled the tents. Everyone was shoving their gear into their backpacks as I lit the little stove. "Today we're having old-fashioned rolled oats with raisins. It will be good fuel for our bodies."

The little stove sputtered and burned slowly. Ten minutes passed and still the water had not come to a boil. Twenty minutes passed. Still no steam appeared, but a weak flame continued to flicker. Exasperated sighs wafted across me.

"Mom, this is stupid," Ben said. "Can't you put on a new fuel canister?"

"It's dangerous to unscrew the canister from the stove until it's empty." I remembered the manufacturer's warnings. "We'll put on a new canister when this one is empty."

Buying a tiny stove with pressurized, self-contained fuel canisters seemed like a great idea during the planning stages of our hike. We would not have to carry loose fuel sloshing around in a gas can. Everything would be sealed up and safe. Now we were learning that when the fuel canister is nearly empty, it produces less and less heat.

"Let's remember to let a nearly-empty canister burn out at the end of a meal." I stooped to examine the flickering flame. "For now, we are stymied."

"Let's build a fire." Joel gathered twigs.

By the time our fire was smoking, the Boy Scouts ate, packed up, and hiked off with a final wave. The friendly, most likely one-sided competition was over. We admitted defeat. We wouldn't see the Boy Scouts again.

Our confident, trail-savvy swaggers disappeared.

Finally, we set the kettle on a little campfire and before long had oatmeal. Our bodies were refueled, our minds had settled down, and we were ready for the day.

Vignette 28: Falling Down and Getting Up
Tuesday, June 12, Day Six

The Lord lifts up all who fall, and he supports all who are bowed down.

(Psalm 145:14 EHV).

It was 9 a.m. when we hiked away from our fairytale campground in Shubert's Gap. It was 9:10 a.m. when Caleb tripped on a root and landed heavily on a rock with one knee.

Falling onto a rock with your knobby knee is one thing. Crashing onto a rock with a thirty-pound pack on is a whole new realm of pain. The pack catapults you down to a debilitating blow. Caleb was writhing in definite pain. Visions of a shattered kneecap raced through my motherly brain.

"Hold still, Caleb," I said. "We'll help you."

In a united family effort, we unbuckled Caleb's pack and pulled it off him, then lifted him into a sitting position. Helpful hands brushed him off and patted his back in empathy. We crowded around Caleb, analyzing the next move, wondering if we even should move.

His knee looked okay, so I gently rubbed it.

"Can you wiggle your toes?" Tom asked.

"Yep," Caleb groaned.

"Can you bend your leg?" I lifted his leg an inch.

"Yep." Caleb demonstrated.

"Let's see if you can put weight on it," Tom said, as he and Ben pulled him up to a standing position.

"Can you walk?" I asked, and Caleb gingerly took a few steps.

"I think you might be able to walk this off," Tom said. "Walking will keep it from tightening up."

"Okay." Caleb's voice held a hint of doubt. He reluctantly swung on his pack.

And with that, we limped down the trail, matching Caleb's pace, hovering close to help avoid another tumble.

We fall down. God brushes us off and lifts us up.

Vignette 29: The Alphabet Game
Tuesday, June 12, Day Six

Then our mouth was filled with laughter, and our tongues with shouts of joy

(Psalm 126:2 ESV).

"This is boring." Joel nudged his pack to a more comfortable place on his back. Even though we were on top of a mountain, there were no views as we continued to lope along.

"Why can't we do something fun?" Ben kicked a stone into the brush.

Full-scale grumbling ensued. I wanted to join in but remembered I was the mom.

"While we walk, let's play the Alphabet Game." A few groans met my suggestion.

"Give it a chance," I added. "It will help pass the time. It goes like this: The first person says, 'I'm going to the store to buy apples,' or something that starts with an A. The next person says, 'I'm going to the store to buy apples and bananas,' remembering what the first person said and adding his own B word. Let's see if we can work our way through the entire alphabet."

"I know," said Carla. "May I start?"

"Sure!"

"I'm going to the store to buy asparagus," she said. "Joel, you're next."

"I'm going to the store to buy asparagus and bologna." Joel pointed to Tom.

"I'm going to the store to buy asparagus, bologna, and cookies," Tom said. "Mom's next."

"I'm going to the store to buy asparagus, bologna, cookies, and donuts." I nodded to Ben.

"I'm going to the store to buy asparagus, bologna, cookies, donuts, and eggs."

"Hey, I don't think I can stand talking about food. I'm going to the store to buy aspirin," Caleb said with a limp and a grin. We all giggled. Maybe we were getting a little delirious.

The game lasted through grocery store items, through animals, and through USA cities. It lasted for an hour of hiking.

"What's next?" Carla asked.

"I have to stop." I massaged my forehead. "I can't concentrate on side-stepping rocks and remembering the alphabetized items at the same time. I'm getting a headache."

"You need Caleb's aspirin." More laughter followed Joel's comment.

Our laughter gave us relief from the grind of the AT.

Vignette 30: Forts
Tuesday, June 12, Day Six

Be to me a rock of refuge. Rescue me, O my God, from the hand of the wicked, from the grasp of the unjust and cruel man

(Psalm 71: 3, 4 ESV).

"Our guidebook says we should be coming to Fort Dietrich Snyder soon," Tom said. "It's three miles from where we started this morning."

"Will it be time for a pack and snack break?" Caleb slumped.

"That should be a good time for a break," Tom said. The American history teacher in him added, "I'll want to take some time to look around."

Shortly, we came to the site of the fort, but there was not much to see. Only a memorial stone remained. It read: "1755, Site of Fort Dietrich Snyder—A lookout post to warn of the approach of enemies in the French and Indian War."

This was one of a chain of forts and lookout posts erected as protection against Indian raids. The fort stood here when George Washington was still a young man surveying with the Virginia Militia, and when the colony of Virginia extended all the way to the Mississippi River. It was before the United States of America existed.

The British claimed what is today the eastern USA, and the French claimed much of eastern Canada. As the French moved south and the British moved west, they bumped into each other, and war threatened. The Delaware tribe, once friendly with the scattering of settlers, fell in with the French. The British settlers suddenly found themselves as the targets of brutal raids. Families, innocently caught in the middle, were murdered and scalped, their simple farm buildings burned.

We stood where Fort Dietrich Snyder once stood. We could not hear the traffic on nearby Route 183, but we could hear the eerie silence of the ancient Indian trail that passed over the mountain. We could imagine stealthy steps approaching from all directions, but we saw nothing except trees and rocks. Not a sign of recent civilization encroached on our view.

"This must be how the area looked back then," I whispered.

"Yea, how can you defend yourselves from anybody creeping through the forest?" Tom glanced at the twisted trees and underbrush. "Yet it must have been a comfort to the settlers who knew they could run here for protection if danger threatened."

Colonial fortresses offered limited protection and eventually crumbled away, yet people once looked to them for protection. We have our safe places, too, places like our homes or our churches or our phones to call 911, but they don't offer us any more security than the forts of the 1700s. Only God is our ever-present, indestructible fortress.

Vignette 31: Sweet
Tuesday, June 12, Day Six

How sweet are your words to my taste, sweeter than honey
to my mouth

(Psalm 119:103 ESV)!

We sat on rocks and passed around a box of healthy black licorice. It was the snack of the day, according to my list. As we munched, I held up one black, sticky piece. "This treat is actually healthy. It's sweetened with honey, not sugar. Plus, licorice has been used for centuries to treat sickness. It can help with stomach problems, breathing problems, and stress."

"You mean it's not just candy?" Caleb's voice held a tone of disappointment.

"Not really. And it's important not to eat too much either because, like all good things, too much is not good, especially if you have problems with your heart or you're pregnant."

"No worries there." Ben tossed a piece into the air and nabbed it with his mouth. "Give me more," he hooted.

The elbow jabbing and giggles began.

"And it can even kill bacteria around tooth decay." I ended my lecture with a lame smile.

"Like this." Caleb flashed a sticky, black grin that the other three admired and immediately mimicked.

Tom sat on his rock, watching the antics and shaking his head. When the laughter subsided, he said, "It's really sweet, almost too sweet, and it's really good for us, as well. That's amazing."

He took a swig of water and added, "I love sugar, but I've had enough of this."

"There's a little more left in the box." I shook the box and handed it to Ben. "Let's finish it."

"Yes, ma'am." Ben jammed his hand into the box.

"I want more," Caleb said.

"Me too!" Joel and Carla chimed in.

Vignette 32: Changing and Trusting
Tuesday, June 12, Day Six

Fear the Lord, you his saints, since those who fear him lack nothing. Young lions may lack food and be hungry, but those who seek the Lord do not lack any good thing

(Psalm 34:9, 10 EHV).

Throughout the late morning, we trudged on, seeing nothing but trees, tripping over roots and rocks. We were slow, averaging only one mile per hour. We were bored, hot, and thinking about food. Again.

"When can we stop for lunch?" Ben scowled.

"We've only hiked three miles today," Tom said. "And it's early for lunch."

"Please!" Carla begged.

Lunch did seem like a good diversion, so we dropped our packs along the trail, sprawled out in the grass, and divided up lunch, including raw almonds, peanut butter to dip graham crackers in, and banana chips. I knew it wasn't a feast, but the nuts and peanut butter provided protein with the grahams, and the bananas were a fruit.

"Is this all?" Tom wrinkled his nose. "You know I don't really care for almonds or peanut butter."

"I know." I sighed. "But I was trying to vary our food and still give us nutritious meals. Can you try?"

Tom nibbled at his pile of almonds and dabbed the peanut butter with his graham crackers. "Well, I guess it's not that bad when your other choice is starvation."

"Wow. First you admit that something like licorice is too sweet for your sweet tooth and now you sample nuts and confess that they're not that bad. How you've changed, my dear." With that I flopped back into the grass and weeds for a few minutes of total relaxation.

"And how you've changed too," Tom said. "Where is your tick caution? Where is your fear of bugs and dirt?"

"Dealing with life on the AT, even for six days, has changed us." I rolled to my side. "We've had challenges like thunderstorms and troubles like rattlesnakes, but through it all, we've been safe and blessed."

Vignette 33: Precious Water
Tuesday, June 12, Day Six

As the deer pants for flowing streams, so pants my soul for you, O God

(Psalm 42:1 ESV).

Pure water is a blessing we take for granted in the USA. In many parts of the world, finding water is much harder than hiking a few hundred yards. Since we weren't always sure where our next water source could be, we did appreciate water more on the trail than we have at any other time in our lives.

The afternoon was clear and hot. We were sweaty and sticky. After a couple hours of hiking, we came to a side trail. The guidebook promised that if we took the side trail 200 yards, we'd come to Sand Spring.

"Splashing our faces with sparkly, cool water would be wonderful." I studied the flushed faces staring at me. "And drinking cool water would be refreshing after the warm water in our canteens."

"What if the guidebook is wrong?" Tom asked. "There have been many springs listed that we couldn't find. Then we will have walked 400 yards roundtrip for nothing. Do we want to add fifteen minutes of fruitless walking to our day?"

Hmmm. The question gave us pause. Nobody wanted to add any steps to our day. Everybody wanted to cool off at a spring.

"The side trail looks easy." Ben peered down it.

"And it's shady," Joel added.

Caleb nodded. "Let's do it." We were not disappointed. A low, moss-covered rock wall protected three sides of Sand Spring. The water rippled through the sand and gurgled off into the nearby woods. We splashed in the little stream, washed our faces, pumped cool water into our canteens, guzzled the water down, and totally forgot how hot and miserable we were a few minutes earlier.

But soon, after we had hiked another mile, we were hot and miserable again. That's when we came to another side trail. This one was three-tenths of a mile long and led to Eagles' Nest Shelter and a promised spring.

"We have enough water, so we don't need the spring." Tom nodded at me. "And it's only 3:30 p.m., so we don't need a shelter yet."

"It would be nice to see the shelter," Ben said.

"Seeing it would add more than a half mile to our hike." Tom swiped at a trickle of sweat on his forehead.

"It would be nice to stay in a trail shelter," Joel added. "We haven't yet."

"If we don't come to another water source before nighttime, we won't have enough water for dinner and breakfast." I raised an eyebrow and stared back at Tom. "Water is so important to our health."

The necessity of water won out. We can't live without it. Sometimes we were actually panting for it.

Vignette 34: Construction Power
Tuesday, June 12, Day Six

Then they will speak about the power of your awesome works, and I will tell about your great deeds

(Psalm 145:6 EHV).

To get to Eagles' Nest shelter, we had to take a side trail, then cross a twenty-foot gully. On the AT, we began to expect that if we came to any obstacle, like a gully, we'd have to slide down one side and then scramble up the other. But here a strong plank bridge spanned the gap, so we could stroll over it.

Tom stopped to admire the construction. "How could they do this out here in the wilderness without power tools?" He examined the big timber braces and heavy planks. "The hiking club responsible for this part of the trail did a great job."

Next we focused on the shelter itself. It was three-sided and made from huge logs. Each one was at least twenty-four inches in diameter. We puzzled over how anyone could have hoisted those logs here in the wilderness. Even the roof was framed by big timber. The shelter had a four-foot deep lower floor for sitting. Behind it and about three feet higher was the level for sleeping. It was deep enough to hold a man in a sleeping bag and wide enough to squeeze in about six people.

"Can you believe this?" Tom stepped back to admire the giant-log construction. "I am in awe."

We all paused to admire.

"This will be perfect for our family." My nesting instinct kicked in. "We are here first, so it's ours for the night, right?" The possible varmints who could join us in a three-sided shelter had not yet entered my mind.

"Yep, that's the law of the trail."

"Yay!" the kids shouted. "No tents to set up!"

We scurried to claim the shelter, rolling out sleeping bags, organizing the food, and hanging our packs on the wooden pegs provided on one wall.

When the guys headed down to the spring for water, Carla and I took the short walk to the outhouse. Carla peeked into the little wooden building first.

"Look, Mom, a real toilet seat."

We were happy every time we had a real toilet seat, but this outhouse had much more. Carla climbed up three steep steps to sit on the throne-like seat, then after using the toilet, we followed the instructions on the door. "Take some cedar chips from a wire basket behind the seat and toss the chips into the toilet hole. That constitutes a flush. We hope you enjoyed this new compost privy."

"I did enjoy it." Carla laughed. "It doesn't even smell bad in here."

"This whole place is sweet. I wonder how they carried in all the heavy hardware. It almost seems superhuman, like angels helped them lift everything."

Tom approached. "But we know only men constructed all this. They were smart, hard-working men who used the gigantic trees and rocks that God created. They could not have built this without their God-given brains and muscles."

Only later did we learn that these men were smart enough to use the power of a helicopter during construction.

Vignette 35: Understanding
Tuesday, June 12, Day Six

Great is our Lord, and abundant in power; his understanding is beyond measure. The Lord lifts up the humble

(Psalm 147:5, 6 ESV).

The guys returned with an abundance of water. All canteens, the big kettle, and our plastic collapsible water jug were full.

"Great job," I said. "We need lots. Thanks!"

First we used it in the big kettle to take sponge baths. Everybody had a tiny washcloth, a flimsy hand towel, and a sliver of soap. Modesty can be achieved even in an open camp, but we were beginning to relax our standards, to understand living on the trail.

Next we hand washed smelly shirts, socks, and underwear in the versatile, big kettle. Everyone was responsible for their own laundry.

"Don't worry about getting out all the dirt and stains." I grinned. "Just wash out the sweat, take away the crustiness!"

Perhaps tonight we'd have enough daylight left to at least dry my liner socks. Roadrunner's patient smile that first night flashed through my mind. I had been expounding on the virtues of dry, clean socks. Since then, we'd learned that you seldom have dry socks, or for that matter, dry anything on the Appalachian Trail. Heavy dews greeted us each morning. We'd sweat, sweat, sweat all day, and then sometimes it would rain.

I was starting to get it, to accept, to understand.

As I served one of my least nutritious meals of the hike, I was deflated. How could my family thrive on instant chicken noodle soup, instant butterscotch pudding, and fiber bars?

"Mom, I loved this meal!" exclaimed Ben. Everyone else agreed.

How could that supper be a raving success, and why were the easy, non-nutritious meals always a hit?

I did like the quick cooking and clean-up time. Maybe we could do quick and nutritious. Other hikers had done simpler preparations than I had, and they had time to relax in the evenings. That was something to consider as our hike continued.

I was starting to get it, to accept, to understand.

Sometimes we just don't get it, or we don't want to accept it. Understanding is often far down the trail.

Vignette 36: Inward Bound
Tuesday, June 12, Day Six

He is not impressed by the strength of the horse. He is not pleased with the legs of a man. The Lord is pleased with those who fear him, those who wait for his mercy

(Psalm 147:10, 11 EHV).

The day was ending well. But two more interesting things happened.

After supper, the kids dragged us one hundred yards along the ridge to an overlook. Tom grabbed his binoculars as we left camp, the only time he used them. From the lookout, we could make out a castle-like structure, silhouetted on a distant mountain top. Even with binoculars, we could not tell what it was. But the view filled us with peace. The mountains and valleys were cast in shades of lavender as an occasional caterpillar swung across the landscape.

At dusk, a lone hiker sauntered into camp. He was bearded with a kerchief tied like a sweat-band around his head, holding back his shoulder-length brown hair. He tossed his pack on the ground, plopped down on his rolled sleeping bag, and smiled at us.

We stared back. I worried that we'd need to share our shelter with a stranger. The kids were fixed on his muscular thighs, the size of tree trunks. Tom rose, waiting.

The hiker broke the awkward silence. "Hello! I'm Inward Bound. You must be Swiss Family Wisconsin. I've been reading your entries in the trail registers."

"We are." Tom stepped forward to shake his hand. "How's your hike going?"

"I'm actually doing the AT for the second time. My first time was in 1988. I feel that I can learn a lot about myself as I walk the trail, trying to find out who I am. After I get to the end of the trail in Maine, I'll hike home to Connecticut."

We all gasped.

"You mean you're hiking the entire trail twice in three years, and then walking home?" Tom asked.

"Yep, I try to cover thirty-five miles a day, sometimes with a head lamp into the night. This time I'm doing a yearbook of the thru hikers for 1990, you know, the people hiking the whole trail or major parts of it. I've already taken 500 photos. May I take yours? You qualify as thru hikers even though you're only planning 500 miles."

We lined up for the camera, then with a wave good-bye, Inward Bound threw his pack over his shoulders. "I have to be in Delaware Water Gap on Saturday."

As he disappeared down the trail, we still stood in our lineup.

Joel glanced at me. "Do you think we'll get a copy of the yearbook?"

"Can you believe how many miles he hikes every day?" Caleb exclaimed.

"Did you see his legs?" Ben studied his own thigh muscles, flexing one leg at a time.

"Inward Bound is searching for the answers to life's questions on the trail, within himself," Tom said.

"Like what?" Ben frowned.

Tom turned to Ben. "Oh, like wondering who I am, why I'm here, and what will happen to me when I die. Hiking the trail cannot tell us how important we are to God or how God plans to take us to heaven."

"That's being upward bound." I said.

Vignette 37: Rest Securely
Tuesday, June 12, Day Six

I will bless the Lord, who guides me. Even at night my heart instructs me. I have set the Lord always before me. Because he is at my right hand, I will not be shaken. Therefore my heart is glad, and my whole being rejoices. Even my flesh will dwell securely

(Psalm 16:7-9 EHV).

Before total darkness descended, we had a family devotional and played a game of Yahtzee with the kids, but I knew the hour of reckoning was bearing down.

This would be our first night to sleep in an open-sided shelter. The shelter that looked inviting and safe in the daylight now looked dark and scary. The back corners were cloaked in shadows, and the front side was exposed to the outside world.

"Did anyone check the back corners for spiders?" I asked.

"Yep, no spiders." Joel drew a checkmark in the air.

"I've noticed that the gypsy moth caterpillars have gone to sleep," Tom said. "It's good to know we won't have to worry about them crawling over us at night."

"Yeah." A shudder crept down my spine. "But we could have other creatures lurking around. Before you climb into your sleeping bags, shine your flashlights on the insides and check for spiders and bugs. The bags have been open for awhile. Something could have crept in."

"All clear, Mom." Caleb nestled in. The rest agreed.

After a thorough check, I inched into my sleeping bag and pulled the top snugly around my head.

I reviewed our situation. We had three sides of strong timber and we were on a platform several feet off the ground. We were probably safer from bears than in flimsy tents. That was reassuring. How many little critters could climb up or fly in to visit? Mice, squirrels, spiders, ticks, mosquitoes, ants, and...

I only fretted briefly. Confidence, exhaustion, and the peace of God's watchfulness covered me.

~Distance hiked on day six: 9.2 miles
~Shubert's Gap to Eagles' Nest Shelter

Vignette 38: Careful Steps
Wednesday, June 13, Day Seven

By the Lord the steps of a person are made secure. Then he will delight in his way. Though he falls, he will not stay down, because the Lord holds him by his hand.
(Psalm 37:23, 24 EHV).

A symphony of birds woke us at daybreak. Dozens were perched on the roof of the shelter, making a delightful racket and a noisy alarm clock. We ate a quick, instant, almost-nutritious breakfast and marched out by 7:30 a.m.

It was a milestone day, returning to civilization. Our goal was Port Clinton, Pennsylvania, where we would pick up our first re-supply food box and where we would feast on a truck-stop smorgasbord that was promised in a guidebook.

The morning trail rolled easily along grassy lanes. Late morning, we finished our food supply with crackers, cheese spread, and almonds for lunch. I took a bow, congratulating myself with how well the first week of food planning had turned out. Even though we were short on calories, every meal had been satisfying, at least, for an hour or two.

After lunch, a steep, downward trail replaced our easy morning hike. Some of the decline was at sixty percent and peppered with loose stones. A slip of the foot could turn into a disastrous tumble. The constant downhill pounding hurt my toes, like someone had whacked them with a hammer.

"My toes hurt too." Tom winced with each step. "So do my knees."

Before long, our nimble-footed but impatient kids pulled ahead. Before much longer, I became weary of Tom's snail pace, and pulled ahead too, suddenly finding that I was hiking alone.

"Please, Lord, keep us all safe as we pick our way down, way down, to Port Clinton," I prayed, over and over. As I neared the bottom, I glimpsed the kids, all four of them.

I breathed a sigh of relief, then reworded my prayer. "Please, Lord, keep Tom safe as he deals with sore knees, sore toes, and his huge pack."

The kids played at an old railroad bridge spanning the Schuylkill River. Caleb fished, and the others skipped stones. I sprawled on the deck of the bridge. Every few seconds one of us glanced at the trail.

"How far back was Dad?" Ben squinted up the hill.

"Not far."

After a couple more minutes, Joel asked, "Should we go back and check on Dad?"

"He's just taking his time." I tried to reassure us all, even me.

Another ten minutes passed and then Carla shouted, "There he is!" We all saw him through the trees, near the bottom. A cheer erupted. Tom joined in.

"Wow! That was a challenge." Tom trudged up to us. "I had to watch every step."

Vignette 39: Unload the Burden
Wednesday June 13, Day Seven

Blessed be to the Lord, who daily bears us up; God is our salvation

(Psalm 68:19 ESV).

The little town of Port Clinton was nestled between two mountains. We had scrambled down the one. The other would be tomorrow's challenge. We took narrow streets to the post office to pick up our food shipment. There we found Inward Bound, an old friend now, and another hiker, waiting for the post office to reopen.

The new hiker introduced himself. "They call me Slim Chance because people said I had a slim chance of hiking the whole AT, plus I'm a skinny guy."

I grinned and glanced from their packs to ours. "Your packs are so small."

"We've both learned to travel light," Inward Bound said. "I hike without a stove and mostly eat nuts, seeds, and dried fruit."

I contemplated carrying no stove, no fuel, and no cooking pots. The reduced weight in our packs was appealing, but having soup, pancakes, coffee, and all the other comforting hot foods was more appealing.

Our new friends helped us lighten our packs, provided packing tape, and entertained us with trail stories. They were aghast at our stuff. Onto the sidewalk we tossed the pack tent, a big telephoto camera lens, heavy binoculars, six plastic plates, six forks, five knives, mosquito nets, extra sweaters, Carla's piano book, three recreational reading books, and much more. We filled and addressed two boxes to be shipped home. In a third box, we packed extra fuel canisters, guidebooks, and information for the northern parts of the trail and sent that box ahead to Delaware Water Gap.

Still waiting for the post office, Carla and I peeked into the Peanut Shop, an old-fashioned candy shop next door, and bought chocolate pretzels and sodas for all.

Finally, the post office reopened. We shipped out thirty-six pounds in three boxes, a bargain at $17, and we claimed a box, as well—our resupply of food, shipped before we left home. Everything in the box was in good shape. I divided the food into bags, according to my daily menus list.

"I wish I had included cooking oil in the shipment. We'll have to look for a store." I sat back on my heels. "We need oil to pop corn and make pancakes."

I handed Joel and Carla the food for the next day. Ben, Caleb, and I carried the full-day bags. It was a good system with the three of us always thankful to be one more day into the hike and minus the weight of one more food bag.

"Mom, why don't you carry the food for the following day?" Ben said.

"That's a good idea," Caleb added. "We can handle the extra weight for an extra day no problem, and then your pack will be lighter sooner."

"Thanks, guys." I gave each a grateful jostle.

As we swung our packs on our backs and said good-bye to our new friends, Inward Bound promised, "We'll write to you in the trail registries." And they did, but we never saw them again.

"Well, do you feel that your load is lighter?" I glanced at Tom.

He did a skip and hop to demonstrate. "Yes, I love hiking with a lighter load."

Vignette 40: Hospitality at Helen's
Wednesday, June 13, Day Seven

For your steadfast love is great to the heavens, your faithfulness to the clouds

(Psalm 57:10 ESV).

"Port Clinton Hotel." I read the sign as we approached an old, gray, three-story hotel. We were on our way to the pavilion where hikers are invited to set up camp. "Do you think we could stop for a flush toilet before we get to the pavilion?" I asked.

A chorus of cheers erupted from the kids. Majority rules, and we stepped into an 1800s stagecoach-stop inn. Behind the bar, a few decades younger than the hotel, stood a grandmotherly figure.

"Hello," Tom said. "We're a family of six hikers, and we were wondering if we could use your bathroom?"

"I'm Helen, and I'm seventy-six years old." Her eyes twinkled. "Of course I know you're hikers. I've seen hundreds like you stop in here for my juicy cheeseburgers. You are welcome to use that single bathroom next to the door but not the nice one down the hall by the restaurant."

We used the bathroom according to need. I went first. It was a hole-in-the-wall, single-user bathroom. I squeezed against the sink to close the door and noticed the sign behind the door: "If it's yellow, let it mellow. If it's brown, flush it down." I burst out laughing.

When I squeezed out the door without flushing, my family was sitting at the bar a few feet away, frowning at me. They must have heard me laugh.

"Be sure you follow the directions on the sign behind the door." I laughed again.

Helen sauntered over. "It's true. If we flush too often, we'll clog our septic system. But, please, some of you should flush." Her eyes creased in a grin.

While five of us perched at the bar, waiting for the last one to use the bathroom, a man walked in, sat next to us, and introduced himself as Ralph Fischer.

"Helen," Ralph said, "I'd like to buy these good folks a round of drinks."

The kids' eyes lit up as they ordered sodas. Tom's and my eyes lit up, too, as we ordered cold mugs of the local beer, feeling God's love reaching down to us through Ralph at Helen's Bar.

Vignette 41: More Hospitality at Helen's
Wednesday, June 13, Day Seven

Good things will come to the man who is gracious and lends, who conducts his business with justice

(Psalm 112:5 EHV).

Ralph launched into stories about Port Clinton. We were surprised that in the 1800s it had been a shipping center between the railroads and the Schuylkill Canal.

"My Boy Scout troop actually found a huge anchor in the river. That was the start of our local museum." Ralph glanced at his watch. "Whoa, I have to run, but stop by the house later, and I'll open the museum for you. I live across from the pavilion."

As Ralph ran out the door, we gathered our packs and headed out, too, excited to find the all-you-can-eat buffet at a truck stop a half mile out of town. En route, we dropped our packs at the pavilion. Carrying them another mile roundtrip was not an option. With a spring in our steps, we continued to the edge of town, clinging to the promise in our guidebook.

What we didn't read in the guidebook is that the restaurant closed at 4:00 p.m. We arrived at 4:15 p.m. Shoulders sagged. Lips quivered. Disappointment reigned.

"Hey, let's go back to Helen and try her juicy cheeseburgers," I said. Faces brightened, and we trudged back into town.

Helen served us the biggest, best, juiciest cheeseburgers we'd ever tasted. Kids who never wanted tomatoes and lettuce at home gobbled them up. We ordered ice water and chips. Tom and the boys ordered seconds on the burgers, emptying Helen's fridge.

And while we feasted, another man sat in Ralph's chair, watching the boys devour their sandwiches. He told the boys he was the administrator for a correctional secondary school nearby. "You kids are really lucky to have parents like yours.

The boys grimaced.

"No, really. Taking the time and effort to have this adventure with you is priceless. What my students wouldn't give for this experience."

As he rose to leave, he said, "Hey, Helen, I'm buying this family another bag of chips." He threw a few dollars on the bar, patted Ben's shoulder, and smiled good-bye.

"People are sure nice around here." Ben dug into the new bag of chips, and then he stopped chewing briefly. "Thanks for bringing us on this hike," he added dramatically, without complete conviction but with a grin.

Vignette 42: Days Gone By
Wednesday, June 13, Day Seven

For he knows how we were formed. He remembers that we are dust. As for man, his days are like grass. Like a wildflower he blossoms. Then the wind blows over it, and it is gone, and its place recognizes it no more.

(Psalm 103:14-16 EHV).

Tom, Joel, Carla, and I stayed at the bar for more than an hour while our two hyper teens took off to fish in the Schuylkill River. We were mesmerized by the news on the flickering TV in the corner and by Helen's stories of other hikers, of her historic hotel, and of herself.

"Did you know that this hotel was a great social center during the stagecoach days?" she asked. "It was also a news hub. Everybody gossiped here."

Without taking a breath, she changed the subject, probably thrilled to have a captive audience. We were thrilled to sit.

"Take a look at this hand-braided rug." She pulled it from a nearby closet. "My mom from the Ukraine taught me how to braid with seven strands. I made this one, but I have one that my mom made. I'm giving that one to the Smithsonian. They will want it because this is a dying art."

We nodded and smiled, nodded and smiled. After awhile, Tom slapped the counter. "Well, we better get going. Mr. Fischer promised to open the museum for us."

We scooted off the bar stools, tired and stiff. As we ambled down the front steps, Caleb ran past us, shouting over his shoulder. "I have to get the stringer in the pavilion. We caught a big bass!"

Tom looked at me with a smile and a shake of his head. Only Caleb would run a mile or two to get a stringer.

When Mr. Fischer opened the door to his house, he handed me a small bottle of cooking oil. "I heard you ask Helen if there was a store where you could buy some. We don't have any stores, but my wife wanted to share some oil with you."

"Thank you." I clasped the bottle to my chest. "Our kids will thank you during every bite of their buckwheat pancakes."

As promised, Mr. Fischer opened the museum and guided us to the big anchor his Boy Scout troop had found in the river. The anchor was a symbol of Port Clinton's history. This tiny town had once been a transportation hub. Named for DeWitt Clinton, who masterminded the Erie Canal, Port Clinton had its Schuylkill Canal, established in 1824. It was also home to the Schuylkill River Navigation System, the Pennsylvania Railroad, and the Reading Railroad. High quality coal from the nearby mines was transported through Port Clinton to many communities and even into Philadelphia.

Men from the town worked on the canal or railroads and sometimes slept on the canal boats during the winters. Others lived in the Port Clinton Hotel. The canal continued its operation until just before World War II in 1939. Today the town is diminished to only 300 people.

At dusk, we pulled ourselves away from Mr. Fischer and the flourishing old days of Port Clinton, days that are no more.

Vignette 43: Calling Home before Cell Phones
Wednesday, June 13, Day Seven

God causes the lonely to dwell together as a household

(Psalm 68:6 EHV).

As we approached the river, Ben and Caleb were already reeling it in and throwing their only fish back. Good decision, but I was sorry Caleb had run a two-mile circuit to get the stringer.

"We're phoning home," I said.

Phoning home was a well-known phrase in the 1990s. ET, the famous extraterrestrial, wanted to phone home, to connect with his people.

And we needed to connect with our people too. Having a plan to phone home may be a foreign concept to today's cell phone generation, but then it was common. Unlike in the current world, we had not connected with anybody via a phone for seven days. Our people knew nothing about what we were doing, how we were surviving, what problems we had encountered, or where we really were. Zilch.

The first pay phone booth since we had started the hike was down the street, and it was time to connect. We were on schedule for our hike, so our family and friends were expecting to hear from us.

"Grandpa and Grandma and our friends will be happy to learn we are surviving." I led the way.

The phone booth was three feet by three feet with cloudy, cracked glass walls and a phone that hung precariously from one corner. Of course, we could not all crowd inside, but we tried.

We'd made arrangements before we left civilization that we'd call home with their permission to reverse the charges. That meant we'd only have to insert the initial quarters to connect and everyone we called would accept the charges. We'd pay them back later.

The familiar ring of Mom and Dad's phone came first.

"Hello?" It was evening, and there was no caller ID, so Mom sounded a little hesitant.

"Mom, hi." I fought tears of relief.

"Oh, Janie, hi, we've been really anxious to hear from you," Mom said, with Dad breathing a sigh of relief in the background. "Is everyone okay?"

"We are all fine," I said, as a chorus erupted around me. "Grandma, we miss your chicken dinners. We miss your beds. We miss your hot cross buns."

I flashed a mother's frown at them. Ben added, "And we really miss you."

I jumped in. "This is harder than we thought it would be, but we've met some nice people. It was great to get into Port Clinton today and have some cheeseburgers. We're all fine."

"Oh, that's so good to hear," Mom said. "We're all fine here too."

"That's great, Mom. We don't want to run up a big bill, so we'll talk again next week. Love you."

"Love you too," she said.

Vignette 44: B.C.—Before Cell Phones
Wednesday, June 13, Day Seven

Therefore let everyone who is godly offer prayer to you

(Psalm 32:6 ESV).

I nestled the phone onto its hook, resting my head against the dirty glass wall of the booth. I wanted a chicken dinner and a soft bed too. I wanted to talk to Mom and Dad about our adventures, but I knew our stories would worry them more than comfort.

I peered out the booth door at our grinning, dirt-scuffed kids. "Grandma and Grandpa loved our short call. Thanks for making them feel special."

Our calls to neighbors and friends were similar—short, just the facts, reassurances, and then a return to silence from both sides for another week. We dragged our feet to the pavilion, homesick for our real lives.

Cell phones were emerging in 1990, but they were big, bulky, and cost thousands of dollars. Even if we could have afforded one, we wouldn't have wanted to add its weight to one of our backpacks. It would have been worthless to us. The batteries died within thirty minutes, and the network was not yet functioning well.

Cell phones would have changed our AT experience. Would we have been glum and homesick every day, hearing about the little league baseball scores, the friends' parties, and the ease of life at home? Would we scare people with our stories? What if we were out of range of a cell tower and no calls could have gone out or in, would we all have been desperately trying to connect? Long-distance conversation could have distracted us from a special view or moment with family. The independence and the adventure of being alone in the wilderness would have been gone.

Still, having a cell phone would have made our lives easier. By connecting us with the outside world, a cell phone could have given us a way to call for help or to assure family and friends on a daily basis that we were fine. We could have checked the weather map for approaching storms. Sometimes we could have just walked and talked with an encouraging friend. Maybe we could have surprised the kids with a pizza delivery at the next crossroad.

Well, we didn't have the option of cell phones, but we did have a direct line with God. He was our connection.

~Total miles hiked on day seven: 8.8 (plus extra miles into and around Port Clinton)

~Eagles' Nest Shelter to Port Clinton

Vignette 45: Struggles and Blessings
Thursday, June 14, Day Eight

The Lord gives strength to his people. The Lord blesses his people with peace.
(Psalm 29:11 EHV).

"I feel like I've been sleeping in a turnpike tunnel." Tom stretched the kinks out of his back. "That highway traffic roared past all night."

"I can't believe I slept at all," I muttered in my it's-all-your-fault voice. "What were you thinking? I felt like spiders were crawling over me. It was late when we got back here, but we could have set up the tents."

"That seemed silly when we already had a roof over our heads."

"Well, that's all we had." I glanced at the four kids huddled together, still sleeping. Our six bags were lined up in a row on a cement floor in the middle of an open pavilion, no walls, no windows, only open space, open to anything, big or small, that felt like crawling in.

I shuddered again and turned to breakfast duties while Tom got the kids going. Since I had a cement step to set our stove on, I started our elaborate breakfast choice, buckwheat pancakes. As the pancakes sizzled, Mr. Fischer

tooted and drove by. I silently thanked him and his wife for the cooking oil and began to feel blessed. This little town had been a blessing to us.

Breakfast was our first meal without plates since we had shipped them home. Cutting each pancake into fourths and stacking it in a cup was no problem. Eating them with a spoon was easy too. Tarheel's advice on our first night was also a blessing.

We were packing up faster. By 7:30 a.m., we had mailed Father's Day cards and kid letters to friends, and we were swinging out of town. Tom, of course, credited that early departure with the fact that we didn't need to take down tents. Looking skeptical but keeping silent is sometimes the perfect answer from me. Tom was blessed.

"Did you hear what Helen said about the climb out of town?" Joel walked beside me. "She claimed it was worse than the toe stuffer we hiked down."

"I know. I'm dreading it." I felt sluggish from a half day off, from lack of sleep, and from the heavy pancake breakfast.

But the climb surprised us. A series of switchbacks gave us several overviews of Port Clinton and the Schuylkill River far below. We soon crested the mountain, and nobody was even panting.

"I can't believe that climb was easy," I said. "Increased stamina is a blessing."

As we absorbed the view, Tom pointed to a small town to the right. "Look, there's Hamburg."

"How do you know about Hamburg?" Ben screwed up his face.

"I just heard about it." Tom glared down at Ben.

Nobody else had heard of Hamburg, and Tom said it so abruptly that we all burst out laughing. He was, however, right. Hamburg was mentioned in the guidebook.

As days went by, when a town would be in sight, Tom or someone would say, "Look, there's Hamburg!" We'd all chuckle. Laughter with family—blessed.

Vignette 46: Sheltered
Thursday, June 14, Day Eight

'I would hurry to find a shelter from the raging wind and tempest.'
[or from the caterpillars and their webs]

(Psalm 55:8 ESV).

The gypsy moth caterpillars were still with us, their webs crisscrossing the trail. We brushed the webs off our faces, dodged some swinging caterpillars, swatted at others. The leaves above us disappeared as the caterpillars devoured every one. Late morning, we veered off the trail for a break at the Windsor Furnace Shelter, the site of an early iron factory. Glassy slag still coated the trail.

As we approached, the outside of the shelter looked uninviting with caterpillar webs draped around the roof and outside walls. The shelter was dark with no windows and only a low door. Inside, however, there was a reprieve from the webs and caterpillars.

"Let's have our snack in here." I pulled out cinnamon sugar pastries and pineapple rings. "I need to crash for awhile." I tossed my pack on the floor and sprawled against it. My prissy, careful attitude with my new pack had long disappeared. If my pack got dirty, it didn't matter. If bugs crawled on it, I could brush them off. I was only interested in collapsing.

The kids found the trail register and eagerly scanned the entries, looking for hikers we'd met.

"Hey, our friend Slim Chance spent the night here," Joel said. "Listen to what he wrote, 'It looks like a lovely day with only a light drizzle of caterpillar doo-doo.'"

We all laughed. Yes, there was the constant sound of a light drizzle in the air, but it wasn't rain. It was hard, little balls of caterpillar excrement, bouncing as it fell.

"We can be relieved that the excrement bounces and doesn't splat." Tom grinned. "What should we write in the trail journal?"

"I caught a nine-inch bass in the Schuylkill River!" Caleb said.

"We loved Helen's hamburgers!" Ben patted his stomach.

"The people of Port Clinton were really friendly," I said.

Tom wrote out a few comments. Outside the sunlight was filtered and soft, but the webs floated everywhere. Nobody was in a hurry to venture out.

"I'm happy to have this shelter for a bit." I sighed, resting on my pack-turned-backrest.

Vignette 47: Mountaintop Experience
Thursday, June 14, Day Eight

All you peoples, clap your hands! Shout to God! Sing a loud song! Yes, the Lord Most High is awesome. He is the great King over all the earth

(Psalm 47:1, 2 EHV)!

Our ascent of Blue Mountain was gradual, then it turned steep and rocky. We encouraged each other, saying that the views would be worth it.

At 1 p.m., eight miles into our day, we arrived at Pulpit Rock which stands at 1,582 feet. From it, one can see the Pinnacle to the left, a high point on the Blue Mountain ridge, and in the foreground, the Blue Rocks. It's a magnificent view, according to the guidebook. Well, it would be a magnificent view if one could see far-ther than one hundred yards. The air was thick with mist. We had only a few feet of visibility.

"This is disappointing." Tom sank onto a rock. "We climb and climb and then trees block our view or, like now, we are supposed to have a wonderful view, and we have fog."

We all were disappointed and stood around in a gloomy clump for a few minutes.

"Well, let's keep going." I tugged on Tom's hands. "The fog could lift before we get to the Pinnacle, the best view in Pennsylvania."

After another one and a half hours and two more miles of hiking, past a tower and a rock field and through a cleft in a rock formation, we approached the Pinnacle, elevation 1,635 feet. We were excited as we hiked the eighty yards on the side trail.

The weather had cleared, almost like God had pulled back the curtain of fog to awe us with his creation.

"Yes, this is what I'm talking about." Tom grinned at me as we emerged upon a breath-taking, aerial view of rolling patchwork farmland far below.

"It looks like we're in an airplane!" Joel exclaimed, and we all nodded, silenced by the beauty.

Because of its reputation as one of the best views in Pennsylvania, the Pinnacle can be crowded with people, especially on weekends. Day hikers enjoy a loop trail to Pulpit Rock and the Pinnacle from Hamburg, of all places. But today we had the whole scenic cliff to ourselves.

The rocks on the edge of the cliff were warmed by the sun, so we took off our shoes and socks to dry out from the daily sweat. We settled back for a relaxing hour of soaking up the view and basking in the sunshine. As we nibbled on beef sticks, we were at peace.

Vignette 48: Downer at the Top
Thursday, June 14, Day Eight

Return, O my soul, to your rest; for the Lord has dealt bountifully with you

(Psalm 116:7 ESV).

Well, the Lord had blessed us with a breathtaking view, but long-term relaxation was not included. Even though the guidebook had suggested exploring the cliffs and
two caves in the area, we'd heard warnings of snakes in them.

"Let's relax here where we are safe." I perched
near the edge of the cliff.

Joel avoided the caves but poked around on
the top. Within minutes, he scrambled backward
and pointed into a little crevice only three feet to
my left. "Look! It's a copperhead snake."

In one quick move, I gasped, jumped to my
feet, and squinted into the shaded crevice. Two
beady, cat-like eyes, set into a copper-colored,
diamond-shaped head, stared unblinkingly back.
The brown, reddish-brown pattern along its body
blended in with the rocks, leaves, and shadows,
making it almost invisible.

I gasped. "How did you even see it?"

He shrugged. "I was just looking."

We knew enough about copperhead snakes
to recognize one. They are pit vipers, like rattle-
snakes and water moccasins, and have heat sen-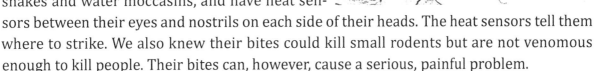
sors between their eyes and nostrils on each side of their heads. The heat sensors tell them
where to strike. We also knew their bites could kill small rodents but are not venomous
enough to kill people. Their bites can, however, cause a serious, painful problem.

Copperheads will more likely strike if they feel threatened, and then they might vibrate
their tails rapidly and give off a musk smell. This copperhead seemed relaxed, but I wasn't
going to sniff for a musk scent.

I glanced to my right. The cliff. I glanced to my left. The copperhead. The saying "Caught
between a rock and a hard place" popped into my mind as I crept away, trying to not scare
the snake.

"Here's another one!" Caleb pointed into a crevice a few feet farther left.

"And here's one too!" Carla backed away from another crack in the nearby rocks. I wondered how many more were lurking, but I didn't want to find out.

"Well, that's enough relaxation for us." I grabbed my shoes and socks and picked my way around all crevices back to our packs. The kids followed, but Tom hadn't budged.

"I'd sure like to admire this view a little longer." But he must have caught sight of the storm cloud on my face and began to pack up too.

"There are some things that you don't mess with," I said, hands on hips. "And venomous snakes are one of them. God protects. But really, don't push it."

I think I heard Tom mumbling, "And don't mess with wives on a rampage either."

The exploring children, the dawdling husband, and the rampaging wife all escaped the Pinnacle without snake bites.

Vignette 49: Upper at the Bottom
Thursday, June 14, Day Eight

He set my feet upon a rock, making my steps secure. He put a new song in my mouth, a song of praise to our God

(Psalm 40:2, 3 ESV).

The over-the-top view and the venomous snake stress at the Pinnacle were contrasted by a relaxing, pastoral end to our hiking day.

Our descent of the mountain was surprisingly rock-free and even followed a dirt road, a real road, for several miles.

"This is too easy for the Appalachian Trail." Tom scanned the area for a trail sign. "Are you sure we're on it? Did it turn up a steep mountain or over a cliff and we just kept moseying along on this easy road?"

We all began to doubt the simplicity of the trail and searched for a white blaze, marking the trail. Finally, a few hundred yards farther, one appeared on a tree.

"Unbelievable. This is so beautiful and so easy it's surreal," I said.

Near the bottom, we stepped into the Garden of Eden. At least, that's what we named it. Lush ferns covered the land, and overhead, trees full of leaves unharmed by the gypsy moths shimmered in the late-day sun.

"Let's walk slowly and breathe in this sweet air," I said.

Of course, the kids had trouble slowing down for long and soon pulled ahead, but still, we all walked softly.

"Look at that colorful parade." Tom nodded ahead. Thigh-deep in ferns, walking single file with various colored packs bouncing on their backs, the kids did look like they were in a parade.

Suddenly the colorful parade halted to watch a timid deer emerge from the trees. It studied them and then us briefly before bounding away.

"Wow!" Ben exclaimed. "Did you see that?"

They spun around to see if we had seen the deer too. Thumbs up!

Thumbs up, God, for this beautiful day of contrasts.

Vignette 50: Happy Wife
Thursday, June 14, Day Eight

You are my hiding place. You will protect me from distress. You will surround me with shouts of deliverance

(Psalm 32:7 EHV).

At 6:15 p.m. we were exhausted and relieved to see the sign for the Eckville Hostel. We found the dilapidated barn with the red door as it was described in the guidebook.

"This looks disgusting." I sniffed in disgust.

The walls were decaying. The windows were broken and, when we peeked inside, we saw a stuffed leg with a boot hanging from the loft. Was it a death threat? I stepped back, frightened, until we read the warning. "This could be you if you try to climb up here."

"Who would want to even step inside?" Ben asked.

Joel peered around the corner. "Here's another sign. It's pointing behind the farmhouse." There we discovered a shed with six wooden bunks, a door, and windows.

"Well, this is more like it." I gave a satisfied nod. "Unlike last night's pavilion, we can close the door on varmints."

Enthusiastically, we unpacked. The kids were unrolling their sleeping bags when I spotted the first monster spider spanning a rafter above me.

"Stop!" I pointed to the spider. Its black, hairy body sported eight legs that extended two inches in every direction.

"Wow." Caleb studied it. "I've never seen such a big spider."

We searched nooks and corners, discovering the first spider had at least two buddies hanging in the rafters.

"They won't bother..." Tom stopped mid-sentence. He saw the storm cloud return to my face. "But what do you think, Janie?"

"No way, no how, neeeveeer will I sleep inside this shed."

"Well, okay, then." Tom shoved things into his pack. "We'll just set up our tents on that nice patch of grass beside the farmhouse."

After the rocky, dusty campsites we'd occupied, this was heavenly. We could even go barefoot on that carpet of green. Across the road, behind the shed, was a solar shower, no warmer than the first one and with the warning to watch out for snakes. We did watch out but loved the chance to shower again. Down the way was a portable privy with toilet paper and beside the farmhouse was a pump that worked. A rickety picnic table and garbage can completed our set up. We had it all.

I simmered the slow but nutritious AT mix for supper, and we snacked on fiber bars during the hour of cooking. Besides showering and doing a little hand-washing, the kids immersed themselves in *Garfield* and *Calvin 'n Hobbs* books that had been stacked in the hostel. Tom and I made notes in our journals.

After supper, clean, well-fed, and relaxed, we climbed into our tents. The tents were on a slight slope, so we felt like we were slipping downhill during the night, but I was happy. I could zip the tent flap shut, barring any monster spiders.

~Distance hiked on day eight: 15.7 miles
~Port Clinton to Eckville Hostel

Vignette 51: Plodding
Friday, June 15, Day Nine

O Lord, my heart is not lifted up; my eyes are not raised too high;
I do not occupy myself with things too great and too marvelous for
me. But I have calmed and quieted my soul
(Psalm 131:1, 2 ESV).

After a simple breakfast of instant oatmeal and high energy bars, we were fired up with high hopes of covering many miles. But Tom's knees were stiff and sore. He lumbered. The rest of us struggled to remain patient with his slow-plod pace.

There were more irritations. The area was again moth infested, and we again took turns leading the group, clearing the caterpillar webs, constantly brushing our faces, arms, and legs free of webs. Even the joy of batting dangling caterpillars at siblings was gone.

The trail was annoying too. It was so rock-filled, we had to pick our steps carefully, moving slower than Tom's initial plod.

And the weather was frustrating. We were encompassed in fog so that the occasional view mentioned in the guidebook was obliterated.

Pain, plodding, moth webs, rocks, fog—we were all irritable.

But what else could we do? So we kept our heads down, eyes on the trail. Then the big diversion of the day happened.

"Did you see that orange lizard?" Carla pointed as one scurried across the trail, barely escaping getting crushed under her shoe. "Oh, it was so cute."

"I think it was really a newt." Joel's encyclopedia brain kicked in.

We all stopped plodding. Newts scurried everywhere. We paused to watch. Carla scooped up newts for potential pets, dropping them into the bottom of her shirt-turned-basket. Joel and Caleb helped. We all oohed and aahed over how precious they were: delicate, three-inch creatures, bright orange color with red spots, and little faces caught in a continuous grin.

Later we learned they were Eastern Red-Spotted Newts and are only found in forests and wetlands of the eastern USA and southeastern Canada. They can live up to fifty years.

Even the newts in Carla's shirt had a chance to live a good, long life.

"We have to release them." I put my arm around her shoulder. "There's no way we can keep them alive on this hike. We don't want them to die."

Carla squatted and gently returned her newfound but beloved pets to their own environment.

We trudged on, spotting a leaping deer and a motionless orange-and-brown turtle to round out the excitement of our day.

After twelve miles of drag-us-down hiking, we came to Pennsylvania 309 and dropped down on a couple of picnic tables outside Gabrinus, a German restaurant. We pulled shoes off weary, wet feet and sat with blank stares for awhile. Done. Exhausted. Quiet.

Vignette 52: Trail Society
Friday, June 15, Day Nine

The Lord is close to the brokenhearted. He saves those whose spirits have been crushed

(Psalm 34:18 EHV).

The delicious smell of high-calorie, fried foods from Gabrinus Restaurant tickled my nose. "I don't think I can walk or cook." I gazed at early-bird arrivers entering the restaurant.

"I've read this restaurant often caters to hikers." Tom took my cue. "If we wash in the bathrooms and put on clean shirts, we should be acceptable."

"What?" Ben perked up. "We're going to eat in a restaurant?"

"We are? Wahoo!" A youthful cheer erupted.

Later, even after our attempts at cleaning up, we looked like a homeless family approaching the hostess, but she smiled and graciously led us to an out-of-the-way table.

Before we had time to order, Tom nudged me. "Look over there. Another hiker is just sitting down. I'm going to say hi."

Sure enough, another scummy-looking character was seated across from us. He must have been right behind us on the trail.

Tom returned, bringing the scummy hiker along. We had the immediate connection of the trail. His hiking name was Boz, and he recognized us from the trail registers as Swiss Family Wisconsin.

"Why don't you pull up a chair and join us?" Tom asked. Boz did.

"So, who's Caleb?" Boz asked. Caleb raised his hand. "I read that you caught a nine-inch bass. Way to go."

"And who jumped over a rattlesnake?"

Again, Caleb raised his hand, and the rest of us laughed.

"But Joel, Carla, and Ben discovered the copperhead snakes at the Pinnacle," I said.

"I can't believe you've seen so many snakes. I've hiked from Georgia and haven't seen one." Boz shook his head. "In fact, I didn't see a view at the Pinnacle, either, because it was raining. You've been blessed."

Like everyone on the trail, Boz looked like a drifter, but he wasn't. He had worked with NASA and was now employed by Lockheed. He'd come alone to hike the trail but had joined up with two other lone hikers. They were farther back and would probably join him the next day.

"Well, I hope we get to meet them," Tom said, and we all agreed.

But now it was time for food. I ordered the salad bar and reloaded my plate several times with the fresh veggies. How often I had longed for a salad along the trail. The kids got plate-sized sandwiches, and Tom gobbled down a swiss steak. Heaping baskets of bread and crackers were served and devoured, served again and devoured. Caleb and Joel even pooled their own money for dessert.

The plentiful food fed our fatigued bodies. Boz's funny and encouraging trail stories boosted our spirits.

Vignette 53: Angels and Bears
Friday, June 15, Day Nine

The angel of the Lord encamps around those who fear him, and delivers them

(Psalm 34:7 ESV).

Outside the restaurant, Boz shook our hands, told us we were doing great, and strolled down the highway to rejoin the AT. He had the relaxed stride of a long-distance hiker. We watched him disappear into the woods and felt the lure to follow. We were refreshed, bodies and spirits, but we had a decision to make.

Tom's knees were hurting, and he thought the anti-inflammatory prescription for his feet would also help his knees, but he needed to refill it.

"If we continue hiking, we'll get to Palmerton late on Saturday or on Sunday," he said. "There's a good chance that no pharmacy will be open. When I paid the restaurant bill, the bartender Scott told me that we could camp in the backyard here, and he would give us a ride into Palmerton in the morning."

"Would we skip the AT from here to Palmerton?" I asked. The thru-hiker mentality was tripping me up. How could we skip fifteen miles of the trail? How could Tom continue fifteen miles with painful knees? Easy decision. He couldn't.

"Camping here is a good idea." I pointed to a grassy patch near a dumpster. "Let's pitch our tents over there." Grass and dumpsters were two of my favorite housekeeping items on the hike.

I played cards with Caleb, Joel, and Carla while Tom and Ben hung our food bag in a nearby tree. When the kids had settled down for the night, Tom and I went into the restaurant bar for a wine cooler and to discuss the morning's ride with Scott.

We'd just leaned back with our first sips when a lady sitting beside Tom asked, "Are you the family camping by the dumpsters?"

We nodded.

"Did you know a bear visits that dumpster almost every night?"

We abandoned our drinks and rushed to the tents. With relief, we saw that the tents were not ripped to shreds.

"Wake up, kids," I whispered. "We're going to move our tents."

The kids grunted, groaned, dragged themselves out of the tents, wanted to know why, but groggily cooperated as we lifted the tents, intact, and moved them one hundred yards. Without further questions and with no answers from us, the kids crawled back into their tents, asleep in seconds. We returned to the bar to thank the lady who had warned us. Our wine coolers were still sitting there, but she was gone.

"Scott, did you see where the lady who was sitting here went?" Tom asked.

"No, can't say that I did, but I've been really busy."

"That's strange," I whispered. "Maybe she was an angel sent to warn us."

Tom raised his eyebrows. "Maybe she was just a nice lady." Tom turned to Scott. "So we'll be ready at 7 a.m. tomorrow?"

Scott nodded as he served another customer, and we grabbed our wine coolers on our way back to the tents. Later, after Tom and I settled into our tent, I tried to stay awake to see the bears visit my beloved dumpster. My vigil lasted ten seconds.

~Distance hiked on day nine: 12.7 miles

~Eckville Hostel to Pennsylvania 309

Vignette 54: Help for the Needy
Saturday, June 16, Day 10

For he delivers the needy when he calls, the poor and him who has no helper. He has pity on the weak and the needy *(Psalm 72:12, 13 ESV)*.

Happy seventeenth anniversary to us! Yes, it was our anniversary, and we were all excited to be going on a ride. We packed up early, saving breakfast for when we got to Palmerton. Scott was ready with his pickup at 7 a.m.

Tom hobbled to the truck. He could not have walked to Palmerton. Scott was an answer to prayers.

Ben and I enjoyed the honor of riding up front with Scott. The rest crammed in the truck bed with all the packs. Scott looked about twenty but was really thirty-ish. He and his mother owned Gabrinus.

"Mom and I recently put the restaurant up for sale," Scott said. "It's time for a change. I've enrolled in college for next semester."

"Good for you," I said. "An education is priceless."

"Yep. I helped my mom and dad with their dreams. It's time for me to follow my own."

As we chatted, I could not ignore the scenery flying by the truck windows. We had moved at a turtle's pace for nine days. It was incredible that now we could be moving so fast. It was also a treat to see houses and remember the comforts of our home.

I thought about carpeted floors, flush toilets, and a normal stove with a garbage can under the kitchen sink. It's not to be, I reminded myself.

As we neared Palmerton, the mountains were stark brown piles of rocks forming a backdrop for the small city. Scott explained that the landscape was barren because pollution from the zinc factories had destroyed the vegetation.

"It looks like an unwelcoming little town," I said.

"Don't judge the residents by the landscape." Scott was more of a prophet than he knew.

He dropped us off on Main Street at 7:45 a.m. While Tom and the kids unloaded the packs, I tried to pay Scott for the ride.

"No way," he said. "I'm just glad to help, and I needed to come to Palmerton one of these days for supplies anyway."

Scott jumped into his truck. "Thank you! Thank you!" we shouted.

Vignette 55: Unplanned
Saturday, June 16, Day 10

Many are the wonders you have done, O Lord my God. No one can explain to you all your thoughts for us. If I try to speak and tell about them, they are too many to count

(Psalm 40:5 EHV).

We had a plan for Palmerton. We were going to stay in the town jail. We'd read that the town jail, located below the city hall, had been converted into a hostel for hikers. What a fun thing to tell our family and friends back home.

But before finding the jail, we needed to eat breakfast.

"Let's head over there and fire up the stove for a big pancake breakfast." I pointed across the street to a city park.

The kids cheered. A day off the trail was already feeling great!

We were swinging our packs onto our backs when we heard a sweet voice with an odd question.

"Are you a hiking family?" A middle-aged redhead was walking by with a friend.

It seemed like an unusual question because we were six dirty people with skinned knees and greasy hair, wearing hiking boots and carrying backpacks.

Tom answered, "Yep, we just caught a ride into town with the owner of Gabrinus Restaurant. I need to get a prescription filled. We're going to stay in town today, relax, and do laundry."

We were an oddity and fascinating to them.

"Where are you staying?" the other lady asked.

"We heard hikers stay in the jailhouse," Tom said. "Do you know where that is?"

"Hey," the redhead said. "Why don't you come home with me? My husband is camping in Canada, and my son is at Fort Bragg. The house is practically empty."

It was her turn to be an oddity and fascinating to us. I glared at Tom, incredulous, trying to convey my distrust of this situation. How could a complete stranger invite a whole family of homeless people into her house after a fifteen-second conversation?

But Tom either missed my signal or ignored it. "Sure!"

Jean, our gracious redheaded hostess, and her friend, Lois, helped us pick up our packs and led us the few blocks to her house. Her house was a double, like so many in the central part of town. It was one building divided down the center into two residences.

"Are you Christians?" Jean asked without hesitation. When we nodded, she continued, "I knew you were, and I'm delighted.

"More people to pray for my daughter Lisa. She's in labor, and the baby is breech. She wants to have the baby naturally, but the doctor said it may have to be delivered C-section. Will you pray for her?"

"Of course, we will," I said. "I had all our babies naturally. Carla was born at home."

"Wow! I know God led us to walk that way this morning." Jean smiled. "We didn't plan to walk Main Street. We never do. And I know He led me to ask you home."

Vignette 56: Hospitality Plus
Saturday, June 16, Day 10

I have been young, and now am old, yet I have not seen the righteous forsaken or his children begging for bread. He is ever lending generously, and his children become a blessing

(Psalm 37:25, 26 ESV).

"Come in! Come in!" Jean held the door open as we slipped off our dusty hiking boots and lined them up on the porch.

"Let me show you around the house." She noticed us standing uncomfortably in the living room, big packs in hand. "Just set them over there." She pointed along the living room wall.

Upstairs she showed us Matt's twin-sized waterbed and another twin bed. "Two kids can have the beds and two can have the floor. It's soft carpeting." All four kids eyed the waterbed.

Back downstairs, Jean led us through the kitchen. "Just feel free to do your cooking here. And over there is the washing machine and dryer. Please use them, and if you want, there are clotheslines outside. Oh, yeah, and the shower is through that door. I'm sure a shower will feel fabulous."

We tried to absorb her hasty directions.

"Do you think you'd want to watch TV?" Jean led us back to the living room. All four kids were nodding. "Here, I'll show you how to turn it on." Saturday morning cartoons lit up the screen, and four kids melted to the floor to watch.

"Just make yourselves at home," Jean repeated. It felt like home, even down to the garbage can under the sink.

The kids only budged from Saturday morning cartoons for their turn in the shower or their turn to have a pancake.

"Oh, I almost forgot to give you this." Jean was heading out on her first errand. "Here's the key to Manny's truck. You're going to have to have transportation to get your prescription filled and other errands."

"Are you sure?" Tom hesitated to take the offered key.

"Of course." She pressed it into Tom's hand.

Jean flew in and out of the house in between going to a hair appointment, arranging flowers for an afternoon wedding, picking up her mother, checking in on her daughter Lisa who was still in labor, and going to the wedding.

When the phone rang, we took the messages. After the first call, I quit trying to explain who I was and just said, "A family friend."

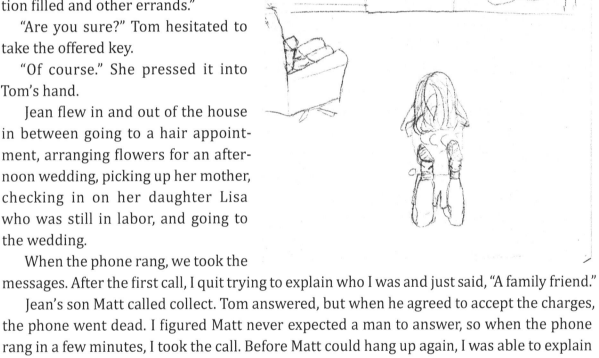

Jean's son Matt called collect. Tom answered, but when he agreed to accept the charges, the phone went dead. I figured Matt never expected a man to answer, so when the phone rang in a few minutes, I took the call. Before Matt could hang up again, I was able to explain how his mom had taken us in. He said, "That sounds like my mom."

Vignette 57: Filling Up
Saturday, June 16, Day 10

Give thanks to the God of Gods . . . He gives food to all living
creatures

(Psalm 136:2, 25 EHV).

Tom dangled the truck key in front of my face and said, "Well, let's take a ride. I need to drop off that prescription, and we can go to the grocery store too."

The kids were glued to the TV.

"Hey, guys, will you be okay if we go to the pharmacy and grocery store?"

The kids were unresponsive, comatose in front of the TV.

"I don't think they'll notice we're gone." Tom led me out the door to Manny's pickup truck.

Having wheels to tool around town was an adrenaline rush. We found the pharmacy and the grocery store, picked up food, including a big bottle of root beer and a gallon of ice cream, and hurried home in case the kids were worried.

They were still watching TV. We turned it off and made a lunch of big hamburgers, macaroni and cheese, lettuce salads, and corn. We wanted to eat more than our stomachs could hold.

Next we all piled into Manny's truck and headed for the Palmerton Community Pool. How often during a hot, sweaty day on the trail did we long for a cool dip? The water was refreshing, and the huge lawn around the pool was soft and warm in the afternoon sun. There were swings on the playground, slides into the pool, and a basketball court that lured Caleb into game after game. We loafed there for hours.

Back at Jean's house, we treated ourselves to root beer floats.

With the kids once more riveted to the TV, Tom and I took off again in Manny's truck. The glory of wheels! First Tom wanted to photograph an old mansion set against the barren mountains. To get the best angle, we went down a ramp and found ourselves heading out of town on a highway. We joked about continuing down the highway, leaving the kids with Jean and having a few days of anniversary escape, but in a couple miles we found a turn-around.

At the pharmacy, we picked up Tom's anti-inflammatory pills and that protection for blisters, moleskin. It had practically replaced all of the skin on Tom's feet.

We also saw a display of root beer and bought more. We had to. After all, how could we finish that gallon of ice cream we had bought earlier without another root beer float?

We ordered a fresh cheese pizza from a local pizzeria, expecting to add meat and condiments before baking but discovered the definition of fresh pizza in Pennsylvania did not mean uncooked. It meant freshly baked. Oh well, the kids were thrilled with a cheese pizza.

Tom even browned some hamburger to sprinkle on top. We finished our feast with straw-berry cake and ice cream. Oh yeah, and more root beer floats.

Our hunger was insatiable. Eating often and a lot helped us bring our calorie count up to survivable levels. Being able to eat as much as one wants is a pleasure for people doing extended hiking. Overeating is a habit that carries into life after the trail. Too often skinny hikers turn into plump people.

Vignette 58: Pure Water
Saturday, June 16, Day 10

You make springs gush forth in the valleys; they flow between the hills

(Psalm 104:10 ESV).

At bedtime, Caleb pulled the long straw for Matt's waterbed. Carla wound up on the other twin bed and Ben and Joel on the floor in their sleeping bags. In seconds, they were in dreamland. The shag-carpeted living room floor beckoned to Tom and me.

But bedtime was delayed by one more chore—cleaning our water pump filter. This filter was supposed to be the best thing on the market for filtering out water impurities, including the giardia parasites found in backcountry streams and lakes. I did not want any of us coming down with an intestinal infection from giardia.

If one or all of us had abdominal cramps, bloating, nausea, and bouts of watery diarrhea, our hike would be over. Even if the infection cleared up by itself in a couple weeks, it could cause ongoing intestinal problems. Prevention was tantamount, and we felt the pump was the best option. We could have treated our water with iodine, but I didn't want to expose all of us to weeks of iodine consumption. We could have boiled our water, but fuel was limited and drinking warm water unappealing.

Striving for sufficient, clean water was life on the trail. When we ran out, we became desperate to stay hydrated, even to survive.

During the first few days of our hike, we pumped all our water. That was seven quarts of water every time, several times a day. The kids took turns pumping, but Ben did the brunt of it, working on his upper body muscles. The pump worked so hard that Ben got blisters on his hands. Every day the pump worked harder.

On day six, Tom reminded me of Tarheel's advice. "If the water flows from a pipe, it's safe." Still, I was squeamish about water flowing from unknown sources out of random pipes projecting from cliffs or hillsides, but when the pump became impossible, we had no choice. That day, as Tom held his canteen under a flowing pipe, he said, "Those men hiked a thousand miles of the trail, and they were doing fine."

Still, I balked.

In Jean's kitchen, with renewed determination to again be cautious, I convinced Tom to help me clean the filter. We used big bowls and a few drops of bleach to sanitize and flush the filter. We struggled with the process, water dripping off our elbows and pooling to the floor. A half hour of mess ensued.

After wiping up the kitchen floor, we checked our efforts. Perhaps the pumped water was a tad purer, which we couldn't tell, but it didn't pump any easier. I shook my head, knowing that we'd continue to trust the piped springs and that our life-sustaining water would continue to be questionable.

I was resolved to that and had snuggled into my sleeping bag when Jean returned from the wedding. An enthusiastic Christian, Jean wanted to discuss her faith. I was drifting, my mind growing hazy. I heard Tom say, "Jesus gives us the living water."

Jean replied, "Jesus says, 'Whoever drinks the water I give him will never thirst. The water I give him will become in him a spring of water welling up to eternal life.'"

Amen, I thought, but I was adrift in a land where pure water bubbled from shiny pipes and impossible pumps were a distant memory.

~Distance hiked on day ten: 0 miles (Distance by pickup: 15.3 miles)

~Pennsylvania 309 to Palmerton

Vignette 59: See You Again
Sunday, June 17, Day 11

You have multiplied, O Lord my God, your wondrous deeds and your thoughts toward us; none can compare with you! I will proclaim and tell of them, yet they are more than can be told (Psalm 40:5 ESV).

"I'm making you breakfast," Jean called into the living room and then up the steps. "It will be ready in twenty minutes."

We'd told Jean we wanted an early start, but we'd all been in deep slumber, the sleep of those whose bodies are worn out. The day was dawning, and another day of hiking loomed. We groaned, stretched, and dragged ourselves from our sleeping bags.

Jean's call did, however, energize us. In twenty minutes, we'd rolled up our sleeping bags, repacked our clean clothes, washed up for breakfast, and gathered around Jean's table.

"I have scrambled eggs, toast, juice, coffee, coffee cake, and cereal." She looked from one eager face to another.

"Jean." I arched an eyebrow. "Are you cooking us our last meal?"

"Very funny! And I hope not. But I did want you to have a good meal before you start your day."

"It will probably be our last good one for a while." Ben stuffed his mouth full of coffee cake.

Later, while everyone helped clean up the kitchen, I packed leftovers from yesterday's shopping spree.

"We'll have one more meal of non-trail food for lunch." I wrapped cheese and lettuce sandwiches, plus added fresh fruit, tootsie rolls, and pizza to my bag.

Jean was hovering nearby. "The AT comes into town along the highways, between bill-boards, and over tracks. It's not a pleasant trail here. Why don't I just drive you to Little Gap where you'll be out in nature, up the mountain, and not fighting traffic."

I glanced at Tom, looking for the affirmative nod. My thru-hiker mentality was again pressed into submission.

"Why not?"

It was 8:20 a.m. when Jean dropped us off near Blue Mountain Ski Resort in Little Gap. "I'll stay with Jean." Carla rolled her lips into a sad face.

She voiced our mutual sentiment, but we hugged Jean good-bye and pushed ourselves toward the trail. Would we ever hear from Jean again, from this faith-filled woman who did wonderful things for us? We waved as she drove down the mountain.

"We will always remember her," Caleb said, sadly.

It was a sad moment. We did not want to say good-bye, and we did not want to begin hiking, but at least we were at the top of the mountain.

Note: We continued to hear from Jean with exchanged Christmas cards, letters, and an occasional phone call. In 2014, Tom and I revisited Palmerton. We again were welcomed warmly by Jean. This time, Manny welcomed us too. We laughed over the many times each of us had told our version of the story. Jean never took in other hikers. We were the first and the last, the best and the worst. We learned how the whole town was abuzz over who was driving Manny's truck that summer day in 1990.

Vignette 60: Pointing the Way
Sunday, June 17, Day 11

I will make you wise. I will instruct you in the way that you should go. I will guide you, keeping my eye on you

(Psalm 32:8 EHV).

We thought we were at the top of the mountain, but we weren't. As we climbed, we found the AT white blazes on rocks. In a few feet, the trail became so steep that the white blazes ahead were at eye level, directly up a rockslide. We were incredulous and even stopped to take a picture.

"Nobody back home will believe this trail," I gasped. The climb was challenging for about one hundred yards and then leveled out, but the rocks did not disappear.

Fog hung heavy, making the air oppressive. Hiking was monotonous, and the cleansing effects of Jean's shower and washing machine evaporated. We made good time, though, and had hiked seven-and-a-half miles by noon.

Since it was Sunday, a few day hikers were out. First we passed a man and two boys and then a man and woman. We greeted each briefly, but when we stopped to sign a register, the man and woman returned to chat.

"Hi, we're Bob and Judy." The man extended his hand to Tom. "Since our cabin is close to the AT, we are curious about hikers. Do you mind if we ask you questions?"

So we answered the usual questions. Where are you from? Where are you hiking? How long is it taking? How are the kids doing? Their curiosity wasn't sated, so they invited us to stop at their cabin.

"We're camping just a few feet off the trail," Judy said. "Why don't you stop in for a break?"

"Probably not." Tom glanced ahead, down the trail.

"We have a flush toilet." Bob attempted to bribe us.

"We just flushed this morning, but thank you," I said.

Then Judy hit our gullible spot. "We have cold sodas."

That did it. Our resolve melted. A hot sun had replaced the humid air. "A cold soda sounds delicious," Tom said.

Bob clapped. "We'll mark the trail where you turn off." They rushed ahead.

We trudged for several minutes through the tall grass and hidden rocks, the hot sun beating down on us, our thirst magnifying.

"I thought they said it was only a couple minutes to their cabin." I fanned my face with my hand. "I bet we missed the side trail. Why didn't we ask how Bob was going to mark it? He probably used some subtle Boy Scout sign that we wouldn't recognize."

My thirst was overpowering. I was longing for that promised soda when we came to a side trail. There on the trail were sticks forming a six-foot arrow, pointing left.

"I guess they really do want us to stop in." I laughed. At another possible fork was another big arrow and at the driveway was another.

"The invitation has certainly been extended," Tom said as we approached a lovely A-frame house.

"They call this camping?" I'd expected a termite-eaten little cabin. "If Bob hadn't pointed the way for us, we'd have never known where to go."

Vignette 61: Rest
Sunday, June 17, Day 11

He who dwells in the shelter of the Most High will abide in the shadow of the Almighty

(Psalm 91:1 ESV).

On Sunday, only a week earlier, our day of rest was less than restful, but today, when we weren't really expecting it, God provided an oasis of rest. Bob and Judy greeted us warmly at the front door and invited us in. We stepped into the tiled foyer. To our left was a spotless white carpet extending into the great room. I threw out my arms, signaling an immediate halt to our family

"Oh, we can't come in with our crusty, dusty boots. And our socks aren't much better."

"Well, okay, then let's sit on the deck," Judy said. "But feel free to use the spare bathroom just inside the back door."

Spare bathroom, as in they had more than one. This was definitely not camping.

The sprawling deck overlooked the valley and beyond to the mountains. We settled into comfy chairs.

"I love chairs with backs." I sighed in pleasure, closing my eyes.

"Don't they all?" I opened my eyes to see Judy's frown.

"On the trail, we sit on logs, rocks, or steps," I explained. "When my back aches, I lie down on the grass or a shelter floor."

"I never thought of that inconvenience," Judy said.

"Oh, there are many other inconveniences, but we'll avoid talking about them." I raised my eyebrows.

Bob emerged from the house carrying a cooler of different flavored sodas. "Help yourselves." He planted himself next to Tom to get acquainted. "We live in Reading, but we built this cabin several years ago to escape the city. We spend as much time as possible here."

While we listened to Bob's story, I handed out our packed lunches, and we settled back to eat sandwiches and guzzle refreshing soda.

Before long, Judy popped out the door, bringing us sliced apples and oranges plus cheese and crackers to complement our lunch. I looked at Ben. "You thought breakfast was our last feast." He nodded with a busted look on his face.

After lunch, Carla went off to play with Bob and Judy's daughter and her friend, both a little younger than Carla. The rest of us lingered over lunch and stories. Too quickly, one-and-a-half hours passed.

"This has been a delightful interlude." I stood up. "But we really must get back on the trail."

Not that we wanted to return to the hot, rocky trail, but we had to hike to reach the Leroy A. Smith shelter four miles ahead. Surprisingly, it was only 4:15 p.m. when we arrived at the shelter. We passed around high fives, impressed with our speed.

The shelter, while web-covered and unappealing on the outside, had a clean, welcoming interior.

"I just need a few minutes to rest my back," I said. "Let's relax."

Vignette 62: AT Support Group
Sunday, June 17, Day 11

Lord, you hear the desire of the afflicted. You strengthen their hearts, and your ear pays attention

(Psalm 10:17 EHV).

Resting was short-lived. I held my foot in the air. "My big toenails have been hurting, but look, the right one is loose."

"That constant downhill pounding does damage." Tom didn't glance at my toe or give me sympathy either; he was distracted by a lone hiker approaching.

Since we were isolated from the rest of the world, I had a momentary flash of fear, but the hiker's friendly face reassured me. Behind him appeared three more hikers. One was Boz.

"Hi, Boz," we yelled.

"Hello, Swiss Family Wisconsin." Boz swung his pack to the ground. "I was surprised from your registry entries that you were ahead of me. At least with you in front, I knew you'd have scared out all the snakes."

Boz introduced us to his friends. Alabama Walking Stick was tall, thin, and from Alabama, of course. He was working on his master's degree in English. Truly Blessed lived in Fort Wayne, Indiana, and worked for the state. We discovered that he ran ultra-marathons near our home. Six Sigmas was a friend of Boz and had just joined them for a week of hiking.

Six Sigmas's hiking name had significance. "Most of the human race is within two sigmas of normal. Six Sigmas is very far removed from normal."

We weren't sure what he meant, but after a few minutes' chat with Six Sigmas, we agreed it was an appropriate name for him. Later we learned that Six Sigmas left the trail the next day. Keeping up with veteran hikers was far removed from normal for him.

Conversation with hikers is like being in a support group. Hikers have endured the same steep climbs, gypsy moth webs, lack of water, and monotonous rocks in the grass.

Our common complaint with these new friends was the sharp, vertical, dinner-plate-sized rocks jamming the trail and camouflaged by ankle high grass. A casual glance at the trail ahead showed only a soft, grassy path, but any attempt to stroll through was an unexpected jolt. Each step had to be deliberate or we risked spraining an ankle or falling onto the rocks. Even our necks had gotten stiff from keeping our heads down, searching for hidden rocks.

The thru hikers said the Pennsylvania AT was the hardest section they'd experienced. That made us feel better. At least we weren't total marshmallows.

While we chatted, Truly Blessed examined his feet and put moleskin on blisters.

"I would think each of your feet would be one big callous by now." I stuck out my moleskin covered right foot. "And look at this. Can you believe I have a loose toenail?"

"Oh yeah, I've lost three toenails already, and blisters never quit."

My shoulders sagged. No sympathy there either, but I was encouraged too, knowing I had a partner in pain, someone who understood. Even for the trivial, God's encouragement comes through the people he puts in our lives.

Vignette 63: Trail Truths
Sunday, June 17, Day 11

The sum of your word is truth, and every one of your righteous
rules endures forever

(Psalm 119:160 ESV).

Boz and his friends refilled their canteens at the spring below and packed up.

"We've heard the next four miles have rocks that are even more treacherous than what we've covered," Boz said. "We want to finish up this section of the trail before dark and be done with it."

As soon as the men disappeared down the trail, we debated whether we should keep going as well. The trail was again luring us. It's true, the trail does beckon. But wisdom, not hiker mentality, played a role in our decision. We stayed.

While the guys set up camp, Carla and I took off for the spring. We found it a half mile down a steep path and then one hundred feet off the trail, secluded by trees and bushes.

The spring was beautiful, water bubbling up through stones into a natural basin and then gurgling down several feet into a pool about twice the size of a bathtub. I was hot and smelly and couldn't resist stripping off my outer layer of clothes for a dip. In seconds, the icy water had me cooled down.

"C'mon in." I splashed water at Carla. She stood there slightly appalled at her mother's immodesty. "It's so cool," I crooned.

Carla loosened up and jumped in too.

By the time we'd rinsed out our shirts, filled our canteens and kettle, and walked back up, up, up to the shelter, we were hot and sweaty again.

Because of our unplanned stop in Palmerton, we had enough food in our supply to have two suppers. We were all ecstatically happy to first have tuna and pasta with granola bars followed by chicken noodle soup and pudding. Hadn't we just had two days of extra food? How true that one's stomach is never totally full on the trail.

As dusk set in, Ben built a pleasant fire to sit around. The gypsy moth caterpillars were active here too. But now, after days of observation, we knew one more truth. Gypsy moth caterpillars settle down at night, and so did we.

Trail truths are not God-inspired and, unlike God's truths, are not indisputable. But talk to a long-distance hiker, and he'll agree. Trail truths are correct, most of the time.

~Distance hiked on day 11: 11.5 miles

~Distance in Jean's pickup truck: 5 miles

~Palmerton–Little Gap to LeRoy A. Smith Shelter

Vignette 64: Lifted Up
Monday, June 18, Day 12

When I said, "My foot has slipped," your mercy, Lord, upheld me.
(Psalm 94:18 EHV).

"Mom, these strawberries taste like fresh." Joel shoveled another bite of his oatmeal topped with brown sugar and strawberries into his mouth.

"Umm." I sampled my own cup of oatmeal. "Can you believe I dehydrated these berries at home?"

It was one of my best breakfasts yet, but we didn't loiter. With one more run on the privy, a three-sided, no-roof little structure, we were hiking by 7:30 a.m., another early, satisfying start.

The trail was not impossibly difficult, as Boz and his friend had suspected. In fact, the rocks seemed better than yesterday. There were many places where the trail was rock free. Another trail truth: Don't always believe what others tell you about the trail ahead.

After about an hour of hiking, Tom stopped. "My left knee is really hurting." While we waited, he wrapped it with an ace bandage.

"Is that better?" I asked, as we continued hiking.

"I think so. I'm grateful this part is easy hiking."

It was on this easy stretch where Ben first fell. He tripped on who-knows-what and sprawled out on the trail ahead. The momentum of the pack pushed him forward, but Ben was not hurt. We picked him up, brushed him off, bandaged a scrape, and continued hiking. A short distance down the trail Ben tripped again and crashed onto the ground.

"I'm okay, really." Ben pulled himself to standing before we could rush over to help. He studied my frown. "Don't worry."

"Ben," I scolded. "You could get seriously hurt. Pick up your feet. Don't shuffle, and watch where you're going."

"Okay, Mom." He gave an apologetic grin and a shrug.

But before long, Ben tripped again. This time, instead of slamming down, he did a sideways somersault, rolling across the bar on the top of his pack and landing off the trail on the side of the hill.

We gasped. He peered up from where he landed in the brush, leaves dangling from his hair, and laughed. We giggled too.

"Lucky you had a roll bar." At Tom's comment, we burst into another round of giggles. We were breathless from laughing and wiping our eyes when a day hiker, wearing a tiny pack, approached from the front. He glanced at us and then at Ben, sprawled beside the trail, still wearing his monster pack.

We struggled to say a friendly hello, but more giggles bubbled up. With a puzzled smile and a shake of his head, the hiker passed by without comment.

"I hope we don't hear stories about a crazy family on the AT," Tom said. "Really, what is so funny about Ben falling?"

We looked at each other, finally getting serious.

"I don't know." I shook my head. "Maybe we just needed the emotional release of laughter. How could that day hiker understand us when we don't even understand us?"

"But you do have to admit, Ben was funny," Joel added, with his little-brother, unsympathetic attitude.

Vignette 65: Paths vs. Highways
Monday, June 18, Day 12

You make known to me the path of life; in your presence there is fullness of joy; at your right hand are pleasures forevermore

(Psalm 16:11 ESV).

Despite delays with Tom's knee and Ben's gymnastics, we still arrived at Lookout Rocks, four miles down the trail, before 10 a.m. From this beautiful spot, we could view the Poconos to the north. In the foreground was Chestnut Valley. Here we took a Tootsie Roll break.

"I love eating candy," Carla said. "I wish we could hike longer so we could have candy every day."

Her three brothers glared at her. Tom and I laughed.

"You want to hike farther?" Ben asked. "I want to quit today."

"Wouldn't it be great if we could get home for part of the summer baseball season?" Caleb asked.

"I could get my paper routes back," Joel added, "and earn enough money to buy a computer sooner."

While the kids dreamed and squabbled, Tom and I soaked up the view.

In Wind Gap far below, cars, trucks, campers, and buses crowded a four-lane highway. We were so far above the highway that the roar of the engines did not penetrate our quiet, but we could see their frantic rushing.

"I wonder where they're all going," Tom said before getting philosophical. "Do you remember what Jesus said about the broad road that leads to destruction?"

"He said, 'Many are on it,'" I answered, quoting a Bible verse I'd learned in Lutheran grade school.

"That highway below is a picture of the wide road going to hell," Tom said. "It's easy and traveled by many. What did he say about the narrow path leading to heaven?"

"'Small is the gate and narrow the road that leads to life, and only a few find it.'" I continued quoting my memory work from Matthew 7.

"This trail is like the narrow path leading to heaven," Tom said. "It's not always easy, and it is traveled by few."

The rest of us agreed that the AT was hard and traveled by few, but at that moment, the thought of sitting in a car and zipping along a highway sounded quite heavenly.

Vignette 66: A Storm to Remember
Monday, June 18, Day 12

> He made the storm be still
>
> (Psalm 107:29 ESV).

Switchbacks down to Wind Gap made the descent easy. Before long, we crossed under Pennsylvania Route 33 and headed up the next mountain. The climb was challenging, and the kids pulled ahead of Tom and me. At the top, their red, sweaty faces reflected our red, sweaty faces.

Caleb scowled. "You're so slow. I should be able to hike ahead of you. I could just meet you at our next camp."

"No," I said, "What if something happens to you?"

"What could happen?"

I attempted diplomacy. "What if we need your help?"

A Wet Day on Wolf Rock   6/18

"Oh, sure." He muttered as he hiked off but stayed within view.

Caleb's grumpiness was contagious, and the stifling humidity made us grumpier. But before our self-pitying attitudes took over, God intervened.

"Was that thunder?" Caleb glanced back at us and skidded to a halt.

It thundered closer. The wind picked up and blew cool air onto our clammy faces. It felt great. We hiked faster.

"Mountain storms move quickly," Tom said. "Should we get out our plastic and cover up with it?"

I studied the clouds. "Nah, I think we can outrun it."

Tom picked the wrong time to believe me. The sky burst open and poured a waterfall down as we scrambled to unpack the plastic and drape it over us. The plastic was creased in folds and our awkward hands couldn't untangle it fast enough. Finally, we were all crowded under the plastic, slightly wet but relieved to be out of the torrents. Rivulets ran off the plastic and into my boots. When I tried to hold the plastic out further, Caleb got wet. We all still wore our packs.

When one moved, his pack pushed into another, but there was no room to take them off and no place to set them to keep dry. Lightning exploded. Thunder boomed. My back ached, my feet hurt, I was shivering, and all we could do was push together under one piece of plastic.

When the deluge weakened to a drizzle, we emerged from our flimsy shelter, checked the sky, and decided to keep hiking. I repacked our plastic.

A day hiker approached us as thunder rumbled closer. "It should be an adventure crossing Wolf Rocks in this. I'll go first and attract the lightning." He sped off.

As another storm moved in, we scrambled once more to pull the plastic out of my pack, open it and dive under. It was roomier. I looked around.

"Where's Dad?"

89

"I'm here." We peered out. Tom stood draped under our little green tarp behind us and gave us the thumbs up.

When the rain slowed to a drip, we started out again. Even though the sky was brighter, this time, I carried the plastic to keep it handy.

Vignette 67: Wolf Rocks
Monday, June 18, Day 12

They were glad when it grew quiet

(Psalm 107:30 EHV).

Before too many steps, we were climbing onto a rocky cliff. "This must be Wolf Rocks." My foot slipped into a crevice. I coaxed my foot out, not hurt but worried. "The rocks are slippery. Be careful, all of you."

Normally the trail would have gone onto a projection of rocks and then turned back, away from the drop off. I studied the area in the logical direction for the trail. No white blaze.

"Over here." Caleb took the lead across the boulders and along the edge of the cliff. Swell. The rocks were slippery and now we had to follow it along the edge of a cliff. I struggled to keep my balance and hold onto the plastic. My heart raced at the thought of one of the kids tumbling off the edge. But they were nimble footed.

Caleb continued to lead, and I hurried the kids along. We'd hiked a couple hundred yards when thunder rumbled closer. It sprinkled, and still the trail did not dip back into the forest but clung to the rocky edge of the mountain.

"Where's Dad?" Ben found himself at the end of our line. This time Dad was nowhere in sight. We called out. No answer.

"You know his knees are hurting more than ever, so he's having trouble keeping up," I said. "Navigating these rocks has to be terrible."

"Do you think he made a wrong turn back at the beginning of Wolf Rocks?" Caleb sounded panicky. "We almost missed it."

It rained harder. I glared at the darkening sky, trying to decide what to do.

"I'll go back." Ben took off, retracing our path.

I spotted a small ravine away from the edge and herded the rest of us into it. We waited and cringed as the lightning flashed and thunder crashed. After several minutes, I dropped my pack. "You three stay here. I'm going to search for them."

I headed back as fast as I could but didn't have far to go. Ben and Tom were only one hundred feet behind us, picking their way across the slippery boulders. Caleb was right. Tom had missed the trail's turn, and Ben had found him.

We regrouped under the plastic and tarp, waiting again for the storm to subside. I peeked out. Mountain laurel bushes in shades of pink and white, forming an archway across the trail in some places, accented the rugged boulders.

Beauty and the Beast. This place was beautiful, but it was beastly right now. When the thunder but not the rain drifted away and we were able to continue, the trail was still hazardous. It zigzagged over the top of the rocks and then dropped steeply. At places we had to sit, scoot forward, and stretch one foot to the next rock below. Each step was a challenge.

When we reached the bottom, it was still raining, but we didn't care about staying dry anymore. It was hopeless. I tucked the plastic into my pack. "Wasn't that smart of me to carry this plastic across the worse trail in Pennsylvania?"

"You should have been a Marine." Tom grinned. "Always prepared."

"I'm just glad the storm is rumbling off, and Wolf Rocks are behind us," I said as heads nodded agreement.

Vignette 68: A Haven
Monday, June 18, Day 12

Then they were glad that the waters were quiet, and he brought them to their desired haven

(Psalm 107:30 ESV).

The trail followed a forest road. Several areas were flooded, but we trudged onward, through the water. Remember, wet is wet, so we were free to splash through all puddles.

The greens were brilliant against the gray rocks and black tree trunks. "Isn't this pretty?" I asked Tom. "When would we have taken a stroll in the woods in the rain?"

"Yep," he said. "I'm sure I wouldn't be doing this anywhere else."

The kids had zoomed ahead but were waiting for us at Pennsylvania 191. "Please may we hitchhike into Delaware Water Gap?" Carla begged and glanced sideways at the boys, the obvious instigators of the question.

Tom and I considered the possibility. We were cold and wet. We even half-heartedly stood by the road, looking miserable and hoping some sympathetic soul would pick us up. But traffic whizzed by.

"It's a little over a half mile to Kirkridge Shelter," Tom said. "Let's go."

"I've read that the hike to Delaware Water Gap is breathtaking," I said. "We wouldn't want to miss that, would we?"

"Humph," was the only reply.

Tom moved slowly, but the rest of us took off like a horse nearing his barn after a long ride. We couldn't get there soon enough. Rain was still falling when we arrived.

Kirkridge was a three-sided shelter with a wooden platform in the middle and dirt floors around the edges. A picnic table was pushed under the roof overhang. The overhang kept most of the rain from blowing into the shelter, but the table was wet.

The kids peeled off wet ponchos and soggy shoes. We lined up our dripping packs along the edge of the dirt floor.

"I'm going to get water for supper." I followed the blue blaze trail. At the top of the ridge, beside a road, I found a water faucet.

I paused to watch the activity across the road at a retreat center. People were jumping into cars and leaving or pulling in with cars and going into the cabins. I longed to be one of them. Some of the cabins glowed with light. I pictured them, warm, dry, and comfortable. I crept closer, contemplating whether I should beg for a better shelter for us. But I talked myself out of it. "We don't have a reservation. We can't just barge in. We don't have extra money for staying in a retreat house. We're fine."

When I got back to the shelter, the kids were indeed fine, wearing dry clothes and sitting on their rolled-out sleeping bags.

Vignette 69: Our Haven
Monday, June 18, Day 12

He brought them to their desired haven. Let them thank the Lord for his steadfast love, for his wondrous works to the children of man

(*Psalm 107:30, 31 ESV*)!

Tom arrived. "Wow, am I thankful to be with you again. That was a really long half mile. I wasn't sure I was going to make it."

I wondered how I was going to make it much longer on the AT with kids sprinting ahead and a husband lagging behind, but there was no time to stew. I needed to make stew.

I heated water to boiling and added Hearty Hungarian, a lentil dish, to the kettle and then pulled it off the burner to soften. Next I heated water for hot chocolate, a warming treat. We waited thirty more minutes for the Hearty Hungarian to cook. When it was ready, I dipped it into pita bread and topped it with pecans.

I handed out the pita with finesse. This was a fancy-restaurant meal, and the family noticed.

"Nice meal, Mom," Ben said. "May we have seconds?"

I nodded. For once I could offer seconds.

"Wow, thanks, Mom!" a chorus sounded.

Outside it was gloomy. Rain continued to pummel the roof and fog drifted in, carrying gloom with it.

"I've had enough of this day," Tom said. "Let's go to bed."

Everyone agreed and crawled into their sleeping bags.

"My sleeping bag is wet by my legs." Ben scooted out. "I must not have had the garbage bag tied tightly around it."

I dug into my pack until I found my rain pants. "Here, put these on. I didn't wear them, so they're dry."

Ben wiggled into them and snuggled into his wet sleeping bag. I smiled to myself and wondered how willing he would have been to wear rain pants to bed at home.

My feet were still in wet socks and freezing. I hated to take my last pair of dry socks out of my pack, but I did. How I appreciated slipping them on before crawling into my damp sleeping bag. With my flashlight tucked into one shoe and my glasses in the other, I had only two thoughts flit through my head. First, what comforts could have been ours in one of those retreat cabins only a couple hundred feet away. My second thought was a prayer. "Thank you, Lord, for making this shelter an almost-dry haven for us."

~Distance hiked on day 12: 13.3 miles

~LeRoy A. Smith Shelter to Kirkridge Shelter

Vignette 70: Wookinpanub
Tuesday, June 19, Day 13

The earth, O Lord, is full of your steadfast love
(Psalm 119:64 ESV).

When I awoke, fog still floated in the air, inside and outside, so I burrowed into the warmth of my sleeping bag. I wasn't budging until we had a view during the descent into Delaware Water Gap and was content to curl up, waiting for the fog to lift.

The others were beginning to squirm when a lady from the Kirkridge Retreat Center wandered past the shelter and stopped to chat. It didn't seem to bother her that she had stumbled into our bedroom. She was studying to be a pastor and wanted to know about our lives on the trail, probably digging for an application in one of her sermons.

After she sauntered away, the fog stubbornly settled on the ground. Giving up on sunshine, I rolled out of my bag and launched my long-prep breakfast, the nutritious, well-loved,

buckwheat pancakes. The timing was perfect, because as the last pancake was eaten and all the dishes were washed, the fog lifted, revealing valleys far below us. Who could have imagined this scene last night when we had stumbled in, blinded by the rain?

It was sunny when we hiked away from Kirkridge Shelter at 10 a.m.

Only a few steps down the trail, we met up with a hiker coming from Delaware Water Gap.

"Where you headed?" Tom stopped to chat.

"I'm doing the whole AT, just doing it backwards. Most hikers start at Springer Mountain, Georgia, in the early spring and follow spring into summer as they head north. Their hope is to arrive at Mt. Katahdin in Maine before the snow flies."

"Isn't south to north the logical direction?" I asked.

"Yep, it was freezing in Maine with temps dropping to ten degrees some nights, but I saw a moose every day for two weeks in a row. And now I can hike into late fall to finish."

"We'll watch for your notes in the registries. What trail name are you using?" Tom asked.

"Wookinpanub," the hiker said and then chuckled at our puzzled expressions. "Were you watching Saturday Night Live in 1981?"

No.

"So you probably didn't see Eddie Murphy playing Buckwheat?"

No.

"Eddie sang 'Wookin Pa Nub in all the wong paces.' It's a spoof on Johnny Lee's 'Looking for Love in All the Wrong Places.'"

We were SNL enlightened and laughed with Wookinpanub.

"So are you looking for love, Wookinpanub, in all the wrong places?" I asked.

"Probably. But I have a better chance of meeting someone special hiking against the current. All the hikers are coming toward me."

"Sounds promising." Tom looked skeptical. "How's the hiking ahead for us?"

"Your views hiking into Delaware Water Gap are grand. And the hostel is pleasant."

With that encouragement, we continued north, and Wookinpanub headed south. From north to south, we hoped he might find love in the right place.

Vignette 71: Déjà Vu
Tuesday, June 19, Day 13

Create in me a pure heart, O God. Renew an unwavering spirit within me.

(Psalm 51:10 EHV).

Was it Déjà vu? Had we been here? We had definitely experienced this day before. The trail and scenery had changed, but the situations were unchanged.

Déjà vu number one: Tom limped along on sore knees. The kids grumbled about our slow pace, and I was caught in the middle as the peacekeeper. When we couldn't stand each other anymore, Tom said, "Okay, enough whining. You kids go ahead."

Happily, they took off. Happily, Tom and I basked in the quiet until a distant rumble sounded.

Déjà vu Two: "Was that thunder?" My shoulders sagged.

"Nah, couldn't be. The front moved through last night."

I was reassured but not totally convinced. The air was hot and heavy again, but the hike on a gravel road was momentarily easy. Before long we came upon the kids, sprawled in the middle of the trail, collapsed on their packs, looking miserable.

"Anyone want a Twix bar break?" I got a few smiles. While we munched on our candy, another rumble sounded. Clouds gathered over the next mountain. We finished our snack, gulped some water, and hurried on. The trail climbed slightly, and at the crest of Mt. Minsi, we looked over the valley far below. Interstate 80 meandered along the valley floor, intersecting the Delaware River.

Déjà vu Three (Incomplete): Yes, it was beautiful, and yet we couldn't totally enjoy it. Thunder was sounding louder, and a rocky descent down the exposed face of the mountain awaited us. A black, streaked cloud hovered over stark Mount Tammany to our right.

"I wonder if Boz and the others are getting wet over there," Joel said.

"I wonder if that cloud is moving in our direction." I had a feeling of impending doom. I did not want to replay yesterday's stormy, rocky hike.

Ben studied the sky and with self-assurance predicted, "That storm is going to miss us."

We gave Ben, the weatherman, a tough time. "Oh, sure, Ben, how do you know?"

He didn't recant, and we stayed dry.

More Déjà vu: The trail to the Delaware River had a few treacherous areas, but the breathtaking views were worth it. Near the bottom, we again strolled through a garden-like area with gurgling streams and colorful rhododendron. Even the trail was rock free and made me wonder if this was a hint of the trails ahead. Tomorrow we would leave Pennsylvania where boots go to die and enter New Jersey.

Déjà vu Times Three: The kids had pulled ahead but were waiting for us by Lake Lenape. "Look what we have." Carla ran to us, holding her shirt out like a basket. From a scrambled mass of legs and tails, a dozen orange newts peered out. Joel and Caleb ran up with their shirts harboring more newts.

"We're going to keep these for pets," Joel said.

"Really?" The monotony of the trail was depleting me.

Vignette 72: Heavenly Hostel
Tuesday, June 19, Day 13

Enter his gates with thanksgiving, and his courts with praise! Give thanks to him; bless his name

(Psalm 100:4 ESV)!

Bye-bye déjà vu. We entered the little town of Delaware Water Gap at 2:30 p.m. We were so happy, we would have skipped to the Presbyterian Church of the Mountains if our packs were not weighing us down. The sturdy, brick church was our destination, our home for the night.

"This church opened its doors to AT hikers in 1976," I explained, "and has welcomed hikers ever since. It was the first hostel on the trail. One of its missions is to reach out to hikers."

"The hostel is in the church basement." Tom spotted the sign and led the way to a side door.

"Are we actually going to sleep inside?" Ben gave me a sideways glance. "I'll miss that damp, three-sided shelter and wearing rain pants to bed."

"It will be quite heavenly," I said and whispered, "Hallelujah!"

"Oh, look at this comfortable room." Tom stepped into the living room.

Caleb scanned the room. "It has couches, chairs, tables, and magazines."

"Look back here." Joel stood in another doorway. "Here is a room with seven bunk beds."

"And there's a shower over in that corner." Ben pointed.

"Where's the toilet?" I asked.

"Here it is." Carla pulled open a door near the entrance.

Our exclamations and packs filled the place as we spread out and made ourselves at home. It was quite heavenly to have our packs in the living room and not crowding us in the tiny back room.

One hiker had already claimed a bunk with his pack. We claimed the other six, thinking it was just the right size for us.

While Carla and I unpacked our toiletries for a shower, a couple members of the congregation came in to do maintenance work. One man noticed Carla's legs and asked, "Do you need first aid for her?"

His question startled me, but as I followed his gaze to Carla's scraped, mosquito-bitten, and dirt-streaked legs, I understood. It only takes an hour on the trail for anyone's legs to look like that. If you haven't hiked the AT, you don't know.

"Thanks, but I think most of that will come off in the shower," I said.

It did.

While Carla dressed, I showered, allowing the hot water to wash away sweat, grime, and body aches. It was heavenly.

Vignette 73: Still Heavenly
Tuesday, June 19, Day 13

Enter his gates with thanksgiving, and his courts with praise! Give thanks to him; bless his name

(Psalm 100:4 ESV)!

While the boys showered, Tom, Carla, and I walked the few steps to the post office to pick up our food shipment.

The post mistress pushed our food box through the window.

"Oh, look, Carla." I pointed to the box. "Mrs. Thiesfeldt drew a big smiley face on our box."

"I miss her." Carla traced the smiley face with her finger.

"Me too." Seeing that reminder of home choked me up.

"Okay, let's get going." Tom grabbed the box and swung toward the door.

"Wait a minute." The post mistress shoved a pile of letters through the window. "I have more."

Now my eyes misted more. Mom and Dad, my sisters, a brother-in-law, and three friends had written us. We returned to the hostel and gathered together in the living room to read each one aloud. How wonderful it was to hear from all those thoughtful people back home.

While Tom showered and rested his knees, the kids and I explored. Across from the hostel was a village store. Actually, it was a combination store, café, and living room for the elderly couple who ran it. A thin layer of dust had settled on everything. When I tried to purchase a jug of apple juice from a rack, the lady said, "Oh, no, that's not for sale. It's ours."

Apple juice was the only item that appealed to me. The kids did, however, buy a few post cards, slightly yellowed, and they replenished their candy supply with stale M&Ms.

A few blocks down, we found an outdoor fruit and bakery stand. That was it for stores. I purchased fresh fruit, an apple pie, and Cokes. Those few items came to $14, more than I wanted to spend, but our excitement at the thought of eating it made the purchase worthwhile. On the walk back, we played the "I love" game.

"I love Delaware Water Gap." Carla skipped beside me.

"I love the food we bought," Ben said.

"I love that we have real beds." Caleb gave one thumb up, one down. "But I wish we had a place to fish."

"I love that we have a shower and a toilet," Joel said.

"I love that we have peace and quiet in our own little room in God's house." I nodded my head in the church's direction.

Vignette 74: Hoodlums in the Hostel
Tuesday, June 19, Day 13

But it is God who executes judgment, putting down one and lifting up another

(Psalm 75:7 ESV).

A couple of long-haired, shiftless-looking young men stood in the church parking lot, rolling cigarettes and smoking.

As we approached, I called out, "Hello, are you hikers?" I was hoping for a no.

"Yep, we've already moved into the hostel," one replied. "My dad is inside."

I followed our kids through the entrance and down the steps, muttering. "This is great, just great. We hiked this trail to get away from scum balls and their influence on our kids, and now we're living with them. I bet those guys out there are smoking pot."

Disgruntled at the thought, I stomped into the basement, almost crashing into another hiker. He had a skinny, greasy braid hanging down his back, several rings dangling from his ears, colorful beads around his neck, and a tie-dyed T-shirt.

"Hi, I'm Levi." He tipped his head. "Did you meet my son and his friend outside?"

"Sure did." I turned away. "We'll move our packs out of your way."

As soon as we pushed our scattered backpacks into a tighter pile in one corner, the kids escaped to the back room to relax. I scurried to the picnic table outside.

There I could cook supper, guzzle my beloved Coke, and maintain an unsociable attitude, but Levi followed me. The three newcomers wanted to chat.

"What you cooking?" Levi sat down on the bench. "We just carry canned food."

"That must be heavy." I poured water into the pot and stirred in rice. "I'm making Spanish rice from a box and adding pinto beans which I dried at home. It's filling and a complete protein."

"I wish we'd done better planning." Levi shook his head. "I'm a teacher and didn't have much time between the end of the school year and our departure." He turned to the young men. "This is my son, Jim, and our friend, Dave." They stepped forward to shake my hand. No lingering smell of marijuana clung to them.

One shouldn't pre-judge. My icy attitude was melting.

"Yeah, I teach in a free school where we have no discipline except what the students determine," Levi said. "We have unusual classes, too, like witchcraft."

"Interesting." Frost returned to my attitude.

"I'm a remnant of the flower children of the 60s." Figured.

But just when I had the trio pegged, Jim added, "Isn't it funny? He's a flower child, and I'm an engineer. What a father-son duo."

"Interesting." The variety of hikers on the AT baffled me, most of them not whom they appeared to be.

Vignette 75: A Host in the Hostel
Tuesday, June 19, Day 13

When you open your hand, they are filled with good things

(Psalm 104:28 ESV).

When the threesome headed out for dinner, I gathered my family outside to eat too. While we feasted on our beans and rice, three more hikers, Paula, Jan, and Tony, introduced themselves and tramped past, into the hostel.

"Who would have guessed so many people were behind us on the trail?" I paused, contemplating the flow of people, and added water to the pot for hot beverages.

"Mom, hello, don't you have apple pie to cut?" Ben pushed the pan toward me. I glanced up to see five eager faces and cut the pie into six equal, massive pieces.

As we sipped our coffee or hot chocolate and savored our tender, delicious apple pie, two more hikers, Alice and Mountain Mist, arrived. All five of these hikers could be fulfilling a retirement bucket list.

Tom bolted up. "What are we doing relaxing out here? We have our belongings thrown all over."

We rushed down the steps into the living room. Things were getting tight. Eight people were trying to claim space. We dragged our packs to the back room, apologizing that we hadn't moved them sooner. And we gathered up hand washing that we'd done earlier and hung around the room. All our underwear was on display, but I was grateful that we had a shelter where the fog and rain couldn't penetrate and where our underwear could finish drying.

"Listen!" Tom held up his hand. Thunder rumbled. A flash of lightning lit the windows. Thunder rumbled again. "Aren't we blessed to be inside tonight?"

Thunder cracked closer and two more hikers, Foxy and Billy Lee, blew in. Foxy was in his early 20s, lanky and dark with shifty eyes. In contrast, Billy Lee had white hair and a quick grin. He walked with a bounce in his sixty-year-old feet.

Early evening, everyone rolled out their sleeping bags. I remembered seeing extra foam rubber pads and mattresses on top of our beds and realized they weren't for the Princess with the Pea. Our heavenly hostel could have often been filled to capacity.

"We have extra padding to share with you." I motioned to our family, and we passed out soft bedding. The grateful hikers grabbed the pads and turned the front room into a wall-to-wall bed. Many of the hikers had shared shelters before and were quite comfortable sleeping together in crowded conditions, especially with the rain pouring outside.

Billy Lee rolled under the corner table with his sleeping bag, peeked out between the legs of the table, and said with sincere appreciation, "I have my spot."

~Distance hiked on day 13: 13.3 miles
~Kirkridge Shelter to Delaware Water Gap Hostel

Vignette 76: The Horrible Hostel
Tuesday-Wednesday, June 19-20, Days 13-14

In the evening, weeping comes to stay through the night, but in the morning, there is rejoicing

(Psalm 30:5 EHV).

"Well, good night, everyone." Tom herded us into the back room.

I sighed with relief, glad to shut the door on that jumble of bodies in the living room but skidded to a halt. The lone hiker was lying in his bunk, staring at us.

"Hello," I said, uneasily. "Sorry you have to share this room with all of us."

"It's certainly better than being out there." He nodded toward the living room, his expression serious.

"We're the Niedfeldt family." Tom headed over, his hand outstretched. "My name is Tom. What's yours?"

"Bob." He kept his hand tucked in his sleeping bag. "I hiked in from the north and am waiting for my son to arrive. We'll hike the AT together for a few days."

I stared at him, straining to hear his monotone voice. Bob shifted his eyes and rolled over. I studied his plump form, bulging inside his sleeping bag, and then spied his spotless boots at the foot of his bed. I doubted his story.

As I said goodnight to my family and wiggled into my sleeping bag on a bottom bunk, I wondered if we were sharing our room with a liar.

Or a serial killer.

The old springs on our beds creaked as we settled into sagging positions. Since a sagging mattress requires many tosses and turns, a creaking, irregular cacophony filled the room.

Quiet Bob was not quiet in sleep. All night, his snores echoed throughout the room like a spastic monsoon.

The irregular cacophony of springs and spastic monsoon of Bob's snores obliterated all sleep for me, so I imagined things and worried.

My legs itched, and I squirmed to kick off any bugs that might be biting me. I wondered if the mattresses had lice or bed bugs. Stealthily, I lifted one side of my sleeping bag and switched on my flashlight, trying to surprise any bugs. I repeated this every half hour but never found any.

Next I strained to hear any unusual movement in the room, keeping vigil to protect my family from serial-killer Bob. Then I began to listen for Foxy, that shifty-eyed character in the next room. Foxy could slip in and attack. I wished I had my cooking pot to slam him.

And so the horrible night continued. I was frustrated to tears, but finally, I saw a hint of morning through the little window above my bed.

Vignette 77: Good Advice from the Pros
Wednesday, June 20, Day 14

I will instruct you and teach you in the way you should go; I will counsel you with my eye upon you

(Psalm 32:8 ESV).

The hostel was buzzing early. At 6 a.m., the bucket-list crowd, Paula, Alice, Jan, Tony, Mountain Mist, and Billy Lee were raring to go. By the time we were packed up and ready to walk to a restaurant for an actual dine-out breakfast, the older crowd was hurrying back.

"Wow," Tom said. "What's the rush?"

"We are all heading in different directions." Paula pointed to Alice and Mountain Mist. "The three of us are catching a bus to Stroudsburg to do laundry."

"We've hiked our distance for this year," Jan added. "Tony and I are catching a bus for home."

"And I'm returning to the trail." Billy Lee took off with a skip to his step.

"Be sure to go to the ma and pa restaurant down the way," Paula called after us.

As we walked down the street in the direction Paula indicated, I leaned toward Tom and asked, "Do you think we'll have that much energy in our 60s?"

"I don't have that much energy now." Tom dragged his feet.

After a hot breakfast of eggs, toast, and crispy fried potatoes, we returned to the hostel to find Quiet Bob, Foxy, and the father-son-friend trio lounging. Bob was waiting for his supposed son, but we asked what the others were doing.

"We're expecting money to be sent from home," Levi said.

"I'm sticking around for the Thursday-night potluck here at the church." Foxy rubbed his concave stomach. "It's every Thursday during the summer. You should stay."

"No, we want to keep on schedule." Tom ignored the pleading eyes of four kids.

"Yep, we are excited to get back on the peaceful trail." I shoved a new roll of toilet paper, still in its wrapper, into a side pocket of my pack.

"Don't you know how to do that?" Without waiting for my obvious, stupid answer, Foxy went to the hostel's grab box and pulled out a medium-sized zip-lock bag. "Here, let me show you."

First he ripped out the inner cardboard core of the toilet paper roll and pulled the strand of toilet paper from the middle. Next he put the roll into the plastic bag with the lose strand on top and flattened the roll. He zipped shut the bag and handed it to me with a proud grin. "Now it's handy and rain-proof. When you need it, open the bag and pull from the middle."

"That's great advice," I said. "I'm totally impressed."

Foxy's grin grew. His face turned into the face of a friendly young man, pleased to help.

Good advice came from a scary-turned-friendly fellow, from someone who was a professional in the skill of hiking. Once more, God opened my eyes to see the person underneath the façade, and every time I opened that dry, handy bag, I thought of Foxy.

Vignette 78: Bad Advice from Novices
Wednesday, June 20, Day 14

Put false ways far from me and graciously teach me your law

(Psalm 119:29 ESV)!

"Well, good morning." A kind-faced lady entered the church hostel. "It's unusual to see a whole family here."

"Good morning," Tom said, as the rest of us smiled, nodded, or echoed Tom.

"I'm Becky, a member of the church. It's my turn to clean." She glanced at our ready-to-go packs. "Which way are you heading?"

"North," Tom said. "We're excited to enter New Jersey today and escape Pennsylvania rocks."

Quiet Bob spoke up, "Don't expect anything better. The AT from the visitors' center for Delaware Water Gap to Sunfish Pond is a very rocky and steep part of the trail. My knees were killing me when I got to the bottom."

I glanced at Tom. A frown creased his forehead. I studied Bob with his flabby tummy and pristine boots. I doubted his boots had touched any trail.

"I've never done it myself," Becky said, "but I hear it's one of the most difficult parts of the trail."

Tom's frown deepened. "My knees have been bothering me. I don't think I can handle a climb like that this morning."

Becky nodded. "I've been told there's a longer but easier way. I've never walked it, but if you take the road past the visitors' center and then a side path up to Sunfish Pond, it should be better."

"We'll try it." Tom's frown relaxed into a smile.

By 9:30 a.m., we were leaving town. The AT crossed the Delaware River via the I-80 bridge. Only a small wall separated our sidewalk from the traffic. When the big semi-trucks flew by, the whole bridge bounced, and our hats flapped. Below us was the wide expanse of the Delaware River and towering on both sides were the mountains. It was breath-taking and a little scary.

As we stepped off the bridge, we cheered to enter New Jersey. We had 140 miles of Pennsylvania's AT behind us and only seventy-two miles to hike across New Jersey.

"Hey, you know what this means?" I asked the gang.

Blank stares.

"It means that when we hike only half the distance that we did in Pennsylvania, we will have conquered New Jersey!"

"Wow! That is something." Tom grinned.

After a short walk on the frontage road, we stopped at the visitors' center to confirm directions. The ranger hesitated to answer our questions and gave halting directions. Now we were confused. She brushed her hair from her eyes, got out a map, and tried again. Tom nodded. I was still confused, but I thought Tom had it. We thanked the ranger and stepped outside.

"Okay. Where do we go?" Tom asked.

"I thought you knew." I shoved my hands on my hips with irritation.

"Well, I think I do." Tom rubbed his brow. "Let's go this way."

Vignette 79: Bad Advice Haunts Us
Wednesday, June 20, Day 14

I hate every wrong road

(Psalm 119:128 EHV).

We ignored the grumblings of four kids as we backtracked and turned right at the stop light onto one-way Old Mine Road. I vaguely remembered the ranger mentioning a one-way road and was reassured.

We hiked the road for two hours, often wondering if we were lost. Finally, early afternoon, we came to a side trail and a sign pointing to Sunfish Pond. The ascent was steep but the trail wide and almost rock free. Occasionally, we glimpsed the rooftops of Delaware Water Gap. The trees were sparse, allowing soft grass to sway in the filtered beams of sunlight. Dark giant rocks peppered the landscape.

"Is that a bear?" Carla skidded to a halt and pointed to a distant boulder.

The rested of us squinted in that direction. "No, it's just a rock," I said.

"Didn't we read that there are lots of bears in this park?" Joel glanced at me.

"Yep, so keep looking."

"Lions and tigers and bears! Oh, my," Carla chanted. Her eyes twinkled.

Caleb took over, "Lions and tigers and bears."

"Oh, my!" all four kids yelled.

Of course, we didn't see any bears.

When the climb became steeper, I kept up with the kids while Tom took his time. In two miles, we welcomed the white blazes of the Appalachian Trail once more pointing our way, but we were disappointed to see the familiar ruts and rock-strewn path we thought we'd left in Pennsylvania.

The sign at the junction told us we'd have another mile to Sunfish Pond. "Let's see." Joel studied his watch. "I calculate that we'll arrive at 2:55 p.m."

And we did.

The kids and I gobbled up our fresh fruit and granola bars dipped in peanut butter. While they fished, I soaked my feet in Sunfish Pond. Tom arrived a half hour later to eat a hurried lunch, then flopped back on the ground. "Wouldn't it be nice to stay here?"

"This area is protected." I pointed to the sign. "And we have a few miles to hike to the next camp." I patted Tom's shoulder. "I'm sorry, but we have to keep going."

As we picked our way around the pond, Paula and Alice arrived.

"Hi!" They acted like we were long-lost friends.

"Is this as far as you've gotten today?" Paula asked.

"Yea," Tom shrugged. "We took the longer, easier route."

"We rode the bus to Stroudsberg, did our laundry, rode the bus back to DWG, and hiked the AT to here," Alice said.

"And how was the AT up to here?" I hoped to redeem our turtle pace.

"Oh, nothing worse than we experienced any place in Pennsylvania," Paula said.

"I guess that means we hiked several extra miles for nothing," I felt defeated.

"Yep, we took advice from two inexperienced people." Tom blew out a big breath. "What were we thinking?"

Vignette 80: Walking Together Apart
Wednesday, June 20, Day 14

Behold, how good and pleasant it is when brothers dwell in unity

(Psalm 133:1 ESV)!

The two ladies joined us for the afternoon. Paula kept pace with the kids and me.

"Don't you two stick together?" I asked as we soon left Alice and Tom behind.

"At first we tried to hike together because it's fun to share stories and talk about the trail and the views, but then we discovered that the faster hiker, me, was always feeling held back and the slower hiker, Alice, was always feeling pushed ahead. It was a strain on both of us."

"It's no wonder the six of us are struggling to hike together."

"It has to be an issue, especially with these young sprinters ahead and your husband hobbling behind."

"I hate being stuck in the middle." Frustration crept into my voice.

"Maybe this would work for you. We pick a designated meeting point every mile or so. I usually get a little reading done while I wait for Alice. When she arrives later, we're both happy to see each other."

"That sounds a lot better than haphazardly hoping we catch the kids sometime before nightfall." I shook my head.

Tom was relieved to discover someone who hiked closer to his pace. For a couple miles, Alice and Tom walked and talked about sore knees. She'd had knee problems hiking in past years, but this year her knees were feeling great. As a precaution, she still carried ibuprofen and a knee brace. She offered Tom a pain pill. He tried it and was amazed how quickly he felt relief. Later, at her encouragement, he also tried her knee brace. That, too, helped, so she told him to keep it.

At Paula and Alice's rendezvous points, we regrouped. Sometimes I was with the kids, sometimes I was with one of the ladies, and sometimes I was with Tom. Varying hiking partners gave us, well, a pleasant variety.

The trail meandered on the ridge of the mountains. There were only a few trees, so we could see for miles on both sides. To the left and far below, the Delaware River glistened in the afternoon sun and, to the right, and also far below, hydro-electric cooling ponds graced the valley.

As I hiked with Alice, she pointed out the ripe blueberries on the bushes along the trail. After that my pace slowed, because I was constantly stopping for another handful. I lamented that the kids were ahead and probably not aware of the treats hanging so near, but when we caught up to them, Carla was picking berries with both hands, a big smile lighting up her blue-smeared face. Paula had shown them the berries.

As the afternoon waned, the four kids took the lead, Paula and Alice were next, and Tom and I brought up the rear.

Vignette 81: Walk 150 Miles in My Shoes
Wednesday, June 20, Day 14

In my distress I called to the Lord, and he answered me

(Psalm 120:1 ESV).

As Tom and I hiked together, we had quiet time to reassess. Even with the ibuprofen and the knee brace, Tom grimaced and flinched down the rough, rutted trail.

"This is no improvement over Pennsylvania's AT." I tripped over a rocky outcrop. If the trail looked bleak to me with healthy knees, I could only imagine how Tom viewed it with every step painful.

When we had launched this adventure, I'd wanted a two-month, 500-mile hike as much as Tom did, but after we'd been a few days into the hike, I'd lowered my goals. Tom, however, had continued to insist that we could make it to New Hampshire. Occasionally, at the kids' pleading, I'd suggested shortening our goal, but Tom would not budge in his resolve.

Despite the futility, I ventured to open the conversation again. "Tom, I would be satisfied if we could hike across the Hudson River and then be done."

"I definitely want to get there," Tom said. "Bear Mountain Bridge over the Hudson was the inspiration for doing this hike. That photo in *Reader's Digest* of a hiker crossing it immediately hooked me."

"Me too. It looked like a peaceful escape from the hectic pace of life. In the photo, the hiker strolled across the bridge while frantic traffic flew by. Who would think hiking the AT could be this difficult?"

"And painful," Tom added.

"We did want to stay in that monastery." I paused to glance at Tom. "Remember, we read about it. It's only a few miles beyond the Hudson. The guidebooks say we can each have a bed plus supper and breakfast."

"Yep, we definitely want to get there."

"What do you think? Should we make the monastery our new family goal?" I held my breath.

Tom grimaced as he inched his way down another ledge of the ridge, then his frown relaxed into a grin. "Sounds good to me. I've been praying about it. We've given this hike a good effort, but it's been exhausting, scary, stressful, and painful. I'd like to see the end in sight, and I know the rest of you do too."

Vignette 82: Only 100 More Miles
Wednesday, June 20, Day 14

You turned my mourning into dancing. You removed my sackcloth and clothed me with joy, so that my whole being may make music to you and not be silent

(Psalm 30:11, 12 EHV).

"What? Are you agreeing to reduce our hike?" I exclaimed.

Tom nodded.

"I've gotta tell the kids." I clapped my hands and hurried ahead.

They were sitting on their packs at the designated meeting point. On one side of them, a mountain stream tumbled. On the other side, Mohican Road ran off into a dirt lane. The kids had scouted the area.

"There's a faucet up the road a little way. It will be good for water," Joel said. "Paula and Alice are at the campground farther up, but it costs too much for us."

I smiled at his inherited, thrifty ways.

"Paula said it's only one hundred miles to the Hudson River," Ben said.

"Yeah," Caleb added. "Do you think we could quit there? Can't you talk to Dad?"

"Yeah, can't you?" the rest begged. I shook off my pack, swung it around, and sat on it, deliberately waiting to make my announcement.

"I already asked him, and he said it sounded good."

There was a moment of stunned silence. They'd asked to quit many times before and had always gotten the same answer from their dad. "We can't be quitters," he would say.

"What?" Ben stuttered.

"Dad says, 'Sounds good!'" I repeated and laughed with glee.

A cheer erupted from the four of them. They skipped around each other, jumping up and down with joy. Tom entered the scene, and we joined the dance.

Snatches of chatter filled the air.

"That's only ten more days!"

"We can enjoy some real summer yet!"

"I can see my friends sooner!"

"Maybe I'll be home for my sixteenth birthday!"

"I'll get to play basketball again!"

Most kids would have been appalled at having to hike one hundred miles. Ours had already conquered 150 miles, so hiking only one hundred more was thrilling.

Vignette 83: Empty and Filled
Wednesday, June 20, Day 14

Let your steadfast love comfort me according to your promise to your servant. Let your mercy come to me, that I may live; for your law is my delight

(Psalm 119:76, 77 ESV).

"Let's camp across the brook," Tom said. "See if you can find two rock-free areas big enough to set our tents."

"Hi folks!" A familiar voice greeted us. Mountain Mist, our newfound friend from Delaware Water Gap, strolled in. He pointed to an area farther off the trail where the grass was tall and soft. "How about over here for your tents?" We liked it. Another tent sat between ours and the stream, but the mystery hiker never emerged that evening.

While I prepared supper, the three younger kids put on their swimsuits and tried scooting down the moss-covered rocks in the stream. They laughed and screeched. By the time they were done, they were also clean and refreshed.

We had the unusual supper of malt-o-meal topped by spam chunks in gravy. They gobbled it up in seconds, and the kids looked for more. I tossed each of them a granola bar and turned my attention to cleaning up from supper.

I was depressed that while the kids were playing, I would have enjoyed sitting on the bank watching. Instead I had to cook supper as daylight faded. I was tired, physically, and tired of working so hard to provide food that never completely satisfied anyone. My feet burned from the wool socks, and my back ached from squatting over the ground-level rock that served as our table.

As soon as my chores were done, I crawled into our tent and collapsed on top of the sleeping bags. After a few minutes' rest, I felt a little better. Carla crept into our tent to keep me company.

"Mom, I'll rub your head for you." She kneaded my head and then rubbed my burning feet too. What a kind gesture, what sweet love, to handle feet that are sweaty-smelly from being crammed in hiking boots all day.

"Now I'll rub your head," I said, and she lay back with a contented grin.

"I have a headache too," Ben called through the tent fabric. "I'm going to bed."

Caleb and Joel, with Mountain Mist's help, started a fire and made popcorn. Tom read a devotional for us. We weren't exactly gathered together in a warm, family circle, but we all were able to hear the comforting words.

It rained gently during the night.

~Distance hiked on day 14: 12 miles or so
~Delaware Water Gap Hostel to Mohican Road

Vignette 84: Sure Feet
Thursday, June 21, Day 15

Many are the troubles of the righteous, but the Lord delivers him from them all. He watches over all his bones; not one of them will be broken *(Psalm 34:19-20 EHV)*.

Everyone slept in. It was 7:30 a.m., eleven hours after we'd turned in, when the tents flapped and bodies stirred. The boys tumbled out first.

"Mountain Mist is gone," Caleb exclaimed. "I didn't even get to say good-bye to my new friend."

"The mystery hiker is gone too," Joel said. "I wonder who he was. Maybe he's a runaway."

Tom and I smiled at Joel's imagination, but I also wondered.

While Tom and I put the finishing touches on packing up, the kids climbed on the branches of a nearby fallen tree. They were getting so sure-footed from our daily battle with rocks that I wasn't even concerned when their game took them fifteen feet above the ground, chasing each other wildly across narrow, crooked limbs.

Caleb Hiking

There were no broken limbs on trees or kids as we began our climb into the Kattatinny Mountains at 9:30 a.m.

"We'll meet at Catfish Tower, two miles ahead," Tom announced.

The four grinned at each other and took off. We could hear Ben say, "Isn't this great? We can move at our own pace and, this is the best part, we only have one hundred miles to go!

The kids had covered the rough trail to Catfish Tower by 10:45 a.m. When I approached it at eleven, I could see them in their colorful clothes, perched like parrots on various levels of the fire tower.

"Come on up. It's cool up here with a nice breeze," Ben called down.

But a wire was strung across the bottom step with a sign hanging from it: *Warning: RF Radiation.*

"What's this?" I pointed to the suspicious sign.

Our scientist, Joel, announced, "That's nothing, Mom. Don't worry."

I believed him and swung my leg across the wire. As I climbed the steps, I realized that I had more to worry about than RF radiation. The railings consisted of one narrow metal bar that ran parallel to the stairs about waist high. Above it and below it was nothing but space, lots of space for kids and adults to fall through.

The wind, too, was not exactly Ben's nice breeze. It was more pre-tornado. But I couldn't be a sissy and miss the view, so I pushed myself upward. At the top, I glanced at the rows of mountains disappearing into the horizon, oohed and ahhed while never releasing my white-knuckle grip from the handrail, and was back down to the safety of the ground before Tom arrived at 11:10 a.m.

"Snacks are down here!" I called to the parrots still hanging in the tower. My plan worked. Before long, all the parrots were safely down.

Vignette 85: Disguised Danger
Thursday, June 21, Day 15

May the Lord bless his people with peace

(Psalm 29:11 ESV)!

The trail leading away from the tower was a grassy jeep road. The area looked harmless, and we strolled along, carefree, with Carla in the lead. Suddenly Joel shouted, "Stop! Snake!" It was the urgent command of danger, and we froze.

Joel pointed to the edge of the grassy road, between Carla and Caleb. The rest gasped and pointed. I kept my eyes on the spot. At first, I couldn't see the snake, its coloring blended perfectly with the sun-speckled grass. Then, like a mystery picture that comes into focus, I saw it, another rattler, stretched out in the grass.

Carla was horror-struck by the idea she'd walked right past the rattler and now was separated from the rest of us. She didn't become hysterical, just looked at us with big eyes and ashen face, ready to sprint back.

"Stay there!" Tom and I shouted in panicky unison.

"We're coming to you, Carla." I used my calm voice. "Just stay still."

Caleb continued on the far side of the trail, creeping past the snake. It did not budge. Carla relaxed. She had a buddy, her big brother, with her. The rest of us followed, one by one. The snake stayed motionless until we were all about fifteen feet beyond. Then Caleb tossed a stone at it. It rattled once and slithered into the tall grass.

We fell together in a momentary group hug and then continued down the trail, still rattled.

The incident was enough to keep the kids behind Mom and Dad, at least until our lunch stop beside paved Millbrook-Blairstown Road. While we gnawed on our beef sticks, we rehashed our latest snake encounter.

"I can't believe people can hike the entire AT and never see a snake." Tom shook his head, frowning.

"Aren't we lucky to see so many?" Caleb asked. "I wish we could tell Boz. He didn't see a snake for a thousand miles."

Ben threw a handful of grass in Caleb's direction. "I guess we're lucky as long as the snakes are peace loving."

"Whoever heard of a peace-loving snake?" Caleb hooted with laughter, and we all joined in.

A car zipped by, followed by an orange Schneider Truck.

"I hope no Hardees truck goes by," Joel lamented. "I couldn't stand to see a picture of all that good food on the side of the trailer."

A colorful picture of high-calorie food flashed through my mind as I handed out our 180th granola bar of the hike. We washed down the dry, scratchy things with gulps of water. Our thoughts turned from peace-loving rattlesnakes to juicy cheeseburgers to monotonous granola bars. Just another day on the trail for Swiss Family Wisconsin.

Vignette 86: Small Pleasures
Thursday, June 21, Day 15

Bless the Lord, who satisfies you with good so that your youth is renewed like the eagle's

(Psalm 103:1, 5 ESV).

The soft, swampy trail climbed slowly to a grassy road on a mountain crest. Walking the trail was a breeze. And there was a breeze. We were treated to glimpses of the lake and farmland far below. This stroll was our first small pleasure of the afternoon.

At 2:15 p.m. we came to a hand pump, the site of our second, third, and fourth small pleasures. Ben pumped. As water splashed out, Caleb cupped a handful of water and flung it at Ben.

"Hey, cut that out!"

Joel and Carla cupped and tossed more water at Ben, and all three giggled while Ben sputtered.

"Okay! Okay," Tom said. "Step aside, Ben."

Seeing his chance, Ben flicked water in every direction, and an epic water fight followed.

A water fight like this might have happened at home when the kids were toddlers. In fact, that was the vision I saw. They were toddlers again, splashing each other in a plastic pool in our yard. Darting away, running in for the super water fling. Soon we were all soaking wet and refreshed.

As our adrenalin surge slowed, we found more pleasures in the cold, clear water of that old pump.

"Why don't I make some iced tea in Ben's big canteen?" I dug into the food bag for a two-quart packet. Ben filled his canteen, sprinkled in the tea mix and shook it. Like cowboys in an old Western movie, we passed that canteen around, wiping our mouths with our arms and saying, "Ahh."

"May we do that again?" Carla asked.

"Why not?" And we did.

Before we hiked on, that old pump provided future pleasure too. We drank enough water to carry us down the trail a few miles, and we filled our canteens in case the next water source was dry.

Vignette 87: Surprise Attraction
Thursday, June 21, Day 15

Let the fields be overjoyed, and everything that is in them. Then all the trees of the forest will sing for joy

(Psalm 96:12 EHV).

At 4:15 p.m. we came to a blue-blazed trail that led two miles to Buttermilk Falls, a dependable water source according to our guidebook. It was pretty enough to be pictured on the front of the New York-New Jersey Guidebook.

"Here's the question," Tom said. "Is seeing it worth adding four miles to our hike?"

"We've already hiked eleven miles today." I chewed on my upper lip.

Joel glanced at his watch. "That took us seven hours."

We debated the pros and cons. Did we want to go out of our way? Did we want to start tomorrow with a negative two miles to make up and that being uphill? Were we dumb to miss a sightseeing highlight? Were we too uptight about covering ground to appreciate the ground? Hadn't we seen pretty waterfalls before, and wouldn't we see more ahead?

We opted to bypass the visual feast and pushed toward Brink Shelter, another four miles ahead.

The kids and I took off, pulled along by the lure of a shelter and supper. In a boggy area as we approached a pond, a flurry of splashing erupted. Hundreds of tiny frogs bounded into the safety of the water.

"Wow!" Caleb said. "Let's do that again."

We backtracked and waited out of sight for a few minutes, giving the frogs time to feel safe enough to crawl out. This time as we approached, we were prepared to watch the frog's antics. Hundreds of tiny frogs all around the pond jumped with abandon, leaping over each other, doing frantic dives into the pond.

We lingered there a minute.

"Wasn't that like the frogs were celebrating life?"

I stood still, absorbing God's creation. Trees formed a dense canopy above our heads and the soft soil cushioned our feet. Cool air hung heavily, suspended between the two.

Vignette 88: Our Shade
Thursday, June 21, Day 15

The Lord is your keeper; the Lord is your shade on your right hand. The sun shall not strike you by day

(Psalm 121:5, 6 ESV).

The comforts of the bog evaporated as we entered the burned-out base of Rattlesnake Mountain. The sun beat down with torturous heat, and the dead air sizzled in our nostrils. Charred skeletons of trees reached toward the blazing sky while scraggly ground-cover plants struggled to regain a footing in the soil. We felt like we were passing through a desert and climbed the rocky trail up the desolate mountain in fearful silence. The name, Rattlesnake Mountain, did little to soothe me.

"I'm really hot, Mom," Carla whined.

Carla and I stopped to rest while the boys continued. Soon they were out of sight. Carla and I crept along. Just twenty feet below the mountain top, Carla wilted into a sobbing heap on the narrow trail.

"I can't breathe, Mom." Her face was red, her skin clammy.

I shifted into action, that first-aid course to the rescue again. Wedged between the rocks, I couldn't even take off my pack, so I stood above her, my shadow falling over her. Next I sponged her with cool water and fanned her with my hat.

"Hey," a voice called down. I glanced up. Ben peered over the edge. Two more heads joined his. "Come up here. It's cooler."

I studied Carla's face. It wasn't as red. "Let's get to the top."

Pulling ourselves to the top was like crawling out of a furnace into the cool breeze of a fan in your face. Carla sank to the ground and leaned on her pack. She had a lazy-boy recliner, backpacker style. I sat over her again so that my shadow fell over her.

"Are you boys okay?" I studied each of their faces.

"Yeah, Mom, we're fine."

While Carla rested with her eyes closed, the boys paced.

"Why don't you keep going? We'll be a few minutes behind you."

"Great! We'll see you at the shelter," Ben said. They took off.

Vignette 89: Not Really Alone
Thursday, June 21, Day 15

God is our refuge and strength, a helper who can always be found in times of trouble. That is why we will not fear when the earth dissolves and when the mountains tumble into the heart of the sea *(Psalm 46:1, 2 EHV)*.

We were alone, just Carla and me. Several minutes passed. Carla slept. I perched on my rock, absorbing the tranquility. Below me, the beiges and browns of the rocks and the stark, blackened trees blended together in peaceful silence. A rickrack of bluish-gray mountains disappeared into the horizon to my right. Rugged rocks towered on my left and were strewn across the mountaintop behind me. Occasionally, a cloud floated across the sun, giving me a reprieve from the intense sun.

Even with my concerns with boys barreling ahead and Tom lagging behind and Carla resting with closed eyes, I was becoming lost in the wild beauty. I was drifting into a calm zone when the quiet was shattered by the roar of a motor.

"What's that?" Carla's eyes fluttered open.

I bolted up. "Could it be a dirt bike or a three-wheeler?" The roar intensified, just on the other side of the rock pile beside us.

Carla sat up, her eyes big. We both strained to see between the rocks, catching glimpses of a three-wheeler careening in a sliding, dusty spin. I held my breath, expecting the three-wheeler and its maniac driver to fly over the rocks and land in a bouncing, steaming heap at my feet. Carla grabbed my hand, and we waited for the earth to split, but the roar faded down the mountain.

How could civilization invade our solitude? The contrast was dramatic. While we were fighting for survival in the wilds, a carefree teenager was only a few feet away, flying along on his three-wheeler. As the whine drifted into silence, I was relieved that the stranger hadn't seen us, that the rocks hadn't crashed onto us, and that we were alone.

Vignette 90: Together Again
Thursday, June 21, Day 15

God settles the solitary in a home

(Psalm 68:6 ESV).

Feeling alone was short-lived. There was movement on the trail below.

"Look! It's Dad." Carla waved over the edge. "Hi, Dad!"

With relief, I saw Tom's gold-colored pack bobbing between the rocks.

Tom glanced up and grinned. He, too, seemed relieved to see familiar faces and soon joined us at the top. After he had rested a few minutes, we were ready to start hiking again. It was 5:30, and we still had two-and-a-half miles to go.

Before her bout with overheating, Carla had balked at being stuck with us, her parents, and had often hiked ahead with the boys. Like every youngest child, she was determined to be as smart, as fast, and as strong as her older siblings. From this point on, however, she was usually content to stay with us. She could count on us to be sympathetic to her needs, a response sometimes—most of the time—lacking in big brothers.

Finally, our hiking partners had evolved into the best combination for us. During the remainder of our days on the trail, the three boys continued to hike ahead, meeting us at designated points. Carla and I kept together, occasionally hiking faster than Tom.

The sun was setting, and the air was cooling. We scrambled over big boulders and around rocky ledges, finally spotting Brink Shelter at 7 p.m.

"Hey, it's Mom and Dad," Caleb shouted, as we plodded into camp. "We're glad you're here."

Vignette 91: Repellent
Thursday, June 21, Day 15

The Lord Preserves The Simple; When I Was Brought Low, He Saved Me

(Psalm 116:6 Esv).

Before I had my pack off, I heard the dreaded question, "What's for supper?"

A tired mother would like to take off her pack and sprawl on the ground, giving her back a chance to recuperate, but a tired, dedicated mother notices that it is getting dark and her family is famished. She resolves to make supper.

The shelter was already occupied. Paula, Alice, and Mountain Mist were sprawled in their sleeping bags, grinning out at us. Not far away, in a patch of grass, the mystery tent stood. Once more, its occupant remained inside, out of sight.

"Better get on some insect repellant," Paula said. "Mosquitoes are really bad here."

"Thanks, but I think I'll just put on some long pants," I said. I hate the smell of insect repellant. I also hate rubbing chemicals on my skin.

"You'll be sorry," floated behind me as I slipped off to put on long pants.

Mosquitoes found my exposed skin, zeroing in on my neck and ankles. I swatted them away and set up my kitchen, putting the stove on an old stump and balancing my utensils on an adjacent log, unstable, but the right height. As the water came to a boil in the little kettle, I added an instant gravy packet.

The mosquitoes intensified their attacks just as a rumble sounded in the distance. Frantically, I rummaged through my pack. "Where's my kettle handle? Has anyone seen it?" I sighed, my arms hanging limply at my side. "Oh no, I bet I left it in the grass this morning."

Paula came to my rescue, pulling out a portable handle so that I could lift the hot kettle and lower it onto another stump. While Ben sawed the rest of the beef jerky into chunks and threw it into the gravy, I brought more water to a boil in our big kettle for mashed potatoes.

My frantic efforts to prepare supper did nothing to discourage the mosquitoes. I put on my jacket and flipped up my hood. That only gave the mosquitoes a more defined target. My ankles.

Finally, in desperation I yelled, "Could somebody please rub mosquito repellant on my ankles?" Joel rushed to my rescue.

"This will help, Mom. The mosquitoes only get so close to this stuff, and they bounce away. Watch."

In seconds, another mosquito approached the invisible barrier on my ankles. It started to zero in for the attack, got wind of the repellant, and sprang away like a torpedo in reverse.

I laughed sheepishly. "Sometimes chemicals serve a purpose."

Vignette 92: Bring on the Storm
Thursday, June 21, Day 15

You will not fear the terror of the night

(Psalm 91:5 ESV).

We weren't actually getting complacent about thunderstorms, just well-seasoned in handling them. As the wind whipped in, we whipped into faster action, not even phased by another setback.

"Mom, the syrup leaked into the food bag and then into my backpack." Caleb dropped the sticky plastic bag at my feet and tipped his pack so I could see.

"Perfect timing." I was already washing dishes in the big kettle and used my dish rag to clean up Caleb's backpack and food bag. He dried the dishes. Tom and the rest set up our tents. Ben and Joel untied our sleeping bags and tossed them into the tents.

"Hey, line up your backpacks inside the shelter." Paula scooted closer to Alice. Brigade-style, Ben, Joel, and Caleb did just that.

The rumbles were rolling closer when Ben and Joel climbed into their tent, Carla into ours. As the first drops of rain fell, Tom and Caleb hung the food bag, and I finished picking up the utensils. A feeling of motherly pride settled over me. Everyone had worked together, preparing for the storm without fear.

As the deluge hit, the three of us still outside our tents scrambled into the shelter. Mountain Mist, Paula, and Alice scooted together even more. The floor was eighteen inches off the ground, a good place to be with a waterfall coming off the roof outside.

"Mom," Carla called from the isolation of our tent. "When are you coming in?

"In a few minutes, as soon as the rain slows. Don't worry. We're right here." Carla smiled at us through the screen, spread out the sleeping bags and settled in.

Vignette 93: Snake Tail or Tall Tale
Thursday, June 21, Day 15

I said in my alarm, 'All mankind are liars.'
(Psalm 116:11 ESV).

While we waited for the rain to subside, we swapped stories. Mountain Mist's story was the best.

"Yes sir, there are lots of rattlers in these mountains. I know of a man who loves 'em so much that he collects 'em. When he finds one, he'll pick it up with a stick and talk to it until it calms down, then he drops it in a basket and takes it home."

Mountain Mist paused and grinned. "Teachers in the area invite this man into their classrooms to share his expertise on rattlers. Of course, he loves talking to the students, and he always has a few live snakes in a cage to bring along for a demonstration too."

Caleb, mesmerized by the story, asked. "Do you think he'd come to Wisconsin?"

"Well, I don't know about that," Mountain Mist said. "But where do you think he keeps his snakes when he's not visiting schools?"

"In a shed?"

"In cages?"

"Wrong! Wrong! You're going to find this hard to believe. He keeps them in his kitchen, free to roam."

"Oh, come on." I studied his leathery face for a hint of a smile.

"Yes, ma'am. They're great pets and won't strike if they don't feel threatened. At last count, he had twenty-five little critters slithering between the table legs and hiding under the refrigerator."

"What does his wife think of this?" I asked

"He lives alone, of course." Mountain Mist broke into a smile.

"Whew!" I said. "At least women have common sense."

Paula and Alice chuckled in agreement. Mountain Mist's story lasted as long as the rain. As darkness descended, we ducked into our tents to dream about rattlers under the refrigerator.

~Distance hiked on day 15: 14.5 miles
~Mohican Road to Brink Road Shelter

Vignette 94: Anticipation and Distress
Friday, June 22, Day 16

Answer me when I call, O God of my righteousness! You have given me relief when I was in distress. Be gracious to me and hear my prayer (Psalm 4:1 ESV)!

The next morning as I crawled from my tent, I gaped in surprise. The mystery hiker was a young woman, maybe even a teen. She threw on her pack.

"Good morning," she said, slapping a cap over her blonde hair. The boys peeked out of their tent as she slipped down the trail, out of sight. We looked at each other and shook our heads.

"Wow," Caleb said, "We didn't know she was a she."

"Maybe she was afraid to hike alone and stayed zipped inside her tent," Joel said.

"That could be." Paula nodded. "I'm grateful to have Alice with me. Sometimes I'm grateful for Mountain Mist."

Mountain Mist, Paula, and Alice laughed together. They were packed up, too, but lingered, drinking hot tea, relaxing under the blue skies, their faces dappled by the sun shining through the foliage.

"I reckon it will take us about two hours to get to the bakery." Mountain Mist wiped out his cup and stuck it into a pack pocket.

"What bakery?" Ben focused his full attention on our friends.

"Didn't your folks tell you?" Mountain Mist asked. "Yep, right there on the trail, only three miles away, you'll find a bakery."

"I'm not sure I'd believe him," Caleb whispered to Ben. "Last night after you went to bed, Mountain Mist was telling lots of stories. Most of them were not truthful."

Mountain Mist grinned mischievously as the three of them took off.

"Is there really a bakery?" Ben's eyes were big with longing.

I nodded. "We're going there too. Dad and I were going to surprise you, but this way you can look forward to it." Their excitement was better than the surprise.

We were nearly packed up when Carla pulled me aside. "What do you think this is?" She pulled down the neckline of her T-shirt. I gasped at the red, swollen, egg-sized bite on her chest and called Tom over. Of course, everyone followed. We all stared at the bite.

I didn't dare whisper what I dreaded.

"I think it's just a spider bite," Tom said matter-of-factly. I nodded my support, not sure that was better.

"What if it's a deer tick bite?" Caleb asked, with little tact. "That'll mean Lyme disease."

"Oh, no," Carla moaned, her face filling with terror. "Will I die?"

I glared at Caleb and hugged Carla. "It doesn't look like a deer tick bite. It's oblong, not round. Even if it is, there is medicine to cure it."

Carla calmed down. I appeared to calm down, too, but inside I was fighting panic. What were we doing out here in the wilderness, exposing our kids to all kinds of dangers?

Vignette 95: Too Much of a Good Thing
Friday, June 22, Day 16

For he satisfies the desire of the thirsty, and he fills the desire of the hungry with good things

(Psalm 107:9 EHV).

The hike to Route 206 held aerial views of the countryside. We clipped off miles, first with views of Lake Culver, shimmering in the distance, then with the lake's deep blues sparkling below us. With such rewarding views, we were surprised when our trail began to descend gently. Yesterday, a southbound hiker had warned us of the cliff-like descent, but his warning was his point of view.

By 10:10, Tom, Carla, and I arrived at the bakery. The boys had been there since 9:20 and were longing to eat something fattening. We all were.

I'd been fantasizing about an ice cream float for a couple of days and was thrilled to find a pop cooler and an ice cream freezer. There was also a small deli. I put my fantasy on hold as we planned a real traditional lunch. Our first course was sliced ham or bologna with cheese on hard rolls and a bag of chips. We sat on the side of the gravel parking lot under a shade tree and feasted on food that our taste buds had nearly forgotten. But I was impatiently awaiting my main course.

Tom and I went in alone to pick out dessert—six cinnamon rolls that were Tom's fantasy, a half-gallon of fudge ripple ice cream, two liters of Pepsi because there was no root beer available, and six large plastic glasses.

Back outside, we passed around the glasses. Stepping out of my sacrificial mother role, I served myself first and let Tom organize the rest. I leaned against my pack to enjoy my fantasy-come-true. The creamy soda and ice cream glided down my parched throat. It was as delicious as I had anticipated. Each of the others also served themselves. While the server served, the unserved calculated and voiced what a fair share should be. I closed my ears to the bickering.

By the time the last person was served, I'd downed my float and was in line for seconds.

"Only one-sixth of what is left," Tom warned, raising a challenging brow.

"I know. I know." I only took my share.

Tom, Joel, and Carla were wise. Some of us were stupid and went for thirds. I joined Ben and Caleb in emptying every container.

Vignette 96: No Trespassing
Friday, June 22, Day 16

I am your servant; give me understanding, that I may know your testimonies

(Psalm 119:125 ESV)!

Before returning to the trail, Caleb, Joel, Carla, and I took a ten-minute walk down the highway to a gas station, to the luxuries of a flush toilet and running water. En route, we discovered beautiful Kittatinny Lake lapping against the backyard of the station and only about twenty-five yards from the bakery's parking lot. It had been hidden by houses.

"You mean there was a lake right here, and we could have been fishing the whole time we waited for you?" Caleb stared across the shimmering water.

"Look, there's a beach." Joel pointed down a driveway that ran behind the bakery. "Do you think we could swim there?"

"Yes, we could fish and swim," Caleb added wistfully, "and have fun."

"As if we aren't having enough fun already," I mumbled.

The beach, about the size of a big backyard, was equipped with playground equipment, a raft, and a bath house. "We'll check it out on our way back to the bakery," I said, thinking how good a cool dip would feel followed by a lazy nap on the warm sand. It had more appeal than climbing Sunrise Mountain with my distended, bloated stomach.

But a fence, plastered with several signs reading *No trespassing! Violators will be fined!* dampened our spirits. I glanced around the deserted beach, wondering who would fine us. Perhaps those warnings wouldn't apply to hot, weary hikers. I studied the dark windows of a house set adjacent to the beach. Should I knock on the door and ask permission? I imagined someone peeking out, studying us, ready to zip out with our tickets the minute we dipped one big toe into the cool water. Our money was running short. We had no leeway for fines. My courage for begging had evaporated as well.

Amidst the disappointed pleading of the kids, I shook my head and led us back to the others. "We can't trespass," I said matter-of-factly, hiding my own disappointment.

Vignette 97: In Shape
Friday, June 22, Day 16

I will give thanks to your name, O Lord, for it is good. For he has delivered me from every trouble, and my eye has looked in triumph on my enemies

(Psalm 54:6, 7 ESV).

As we climbed Sunrise Mountain, I fluctuated between trying not to vomit and wishing I could. My heavy, sugary, lunch was tugging me down with every upward step. I wasn't alone. Everyone was panting, wiping sweat, taking extra breaks.

"Let this be a lesson to us yet again." I addressed myself. "Next time we get to a store, we're going to eat sensibly. Have fresh fruit and vegetables. Drink juice. Eat protein." But deep inside, I felt the slumbering food fantasy stir.

By the time we'd climbed the two miles from the road to Culver Tower, we were back in form. With a parking lot, adequate railings, and no radiation warnings, Culver Tower was meant for tourists. We set our packs aside and scrambled to the top for a breezy view.

Lake Kittatinny nestled between the mountains behind us. Ahead, on the farthest ridge, we spotted a tiny pinpoint.

"That must be High Point," Tom said. "It's the highest spot in New Jersey."

I strained to get a better look, marveling that we planned to walk up and down the hazy ridges until we arrived at that pinpoint. When we hiked along, one step at a time, we couldn't visualize the actual distance. I gazed again in disbelief.

Regrouped at the base of the tower, we were donning packs when several red-faced and sweat-streaked teenagers straggled into the area and collapsed in exhaustion on picnic tables or sprawled in the grass.

In his usual, friendly-to-strangers manner, Tom asked, "Where did you hike from?"

"The counselor made us hike all the way from the parking lot down there. See it through the trees?" One boy pointed down one hundred steps.

"Don't miss the view at the top of the tower," Tom said. "It's breezy, and you can see for miles and miles."

"No, thank you. This is as far as I go." The others agreed.

Our own children, with fifteen- to thirty-five-pound packs weighting them down, stared with disdain, incredulous that their peers could be so pathetic.

Ben shook his head and led his brothers off. Before their packs disappeared down the trail, I noticed the spring in each one's step. What a contrast to the red-faced, defeated kids sprawling at my feet. I felt a surge of victory for the challenge we'd placed before our kids, the challenge they'd accepted. Maybe the occasional dangers we faced were worth it in building life-long character.

Vignette 98: Singing and Dancing
Friday, June 22, Day 16

You turned my mourning into dancing. You removed my sackcloth
and clothed me with joy

(Psalm 30:11 EHV).

The day was hot. To take our minds off the heat, Tom taught Carla one of his all-time favorites, direct from the day of Peter, Paul, and Mary.

Have you ever climbed on a big, high mountain
To look down the other side?
And the world unfolds for your eyes to see
And there you stand with a wonderful sort of pride.

It's a land of plenty, a land of beauty, A land that we all can share.
I sing his praise. He's got mighty big ways.
Mighty big ways.

Before long, the three of us were belting the mountainsides with our chorus, not worrying about who could hear us coming. Soon we spied the boys' packs on the side of the trail, a signal that they were somewhere nearby. They were faking sleep on the rocks to our left, telling us subtly that they had been waiting for a long time. Ben and Joel were flattened against boulders. Caleb had his head resting on his pulled-up knees, snoozing siesta-style.

We joined them. The sandy soil and thick carpet of pine needles underfoot, plus the scraggly trees clinging to the sides of the cliff, did give me the feeling that we were somewhere in the Southwest and not in New Jersey. The landscape shimmered in the afternoon heat.

Later, three and a half miles from Culver Tower, the six of us came to another man-made pleasure in Sunrise Mountain Shelter. The huge pavilion sported a cement floor, a big-beamed ceiling, and no walls so the 3 p.m. breeze skimmed through. At its 1600-foot elevation, it provided a beautiful lookout too.

"Let's take another break." Tom propped his pack against a pole. I followed suit and brought out snacks.

In between swinging from the rafters and dancing across the smooth, rock-free floor, the kids joined us for dried pineapple rings and, sigh, granola bars. Water, picnic tables, and the parking lot were not in sight. Those were two-tenths of a mile down the trail. No tourists were venturing that far from their cars. We were all alone and loving it.

Vignette 99: Dirty Desperation
Friday, June 22, Day 16

As a deer pants for flowing streams, so pants my soul for you, O God

(Psalm 42:1 ESV).

Our next meeting point was Mashipacong Shelter, our goal for the day. It was still 3.4 miles ahead and pulled us on. The boys took off, eager to be done for the day.

Carla and I hiked with Tom for awhile, but I began to think about the hand pump our guidebook promised and rushed toward it. Without telling Tom and hardly realizing I was doing it, I led Carla along, obsessed with the refreshing thought of washing up there. Soon we were too far ahead of Tom to bother with waiting for him.

Our pace was frenzied as we stumbled on. I was getting hotter and more desperate with every step.

"It can't be much further." I flicked a drop of sweat off my nose.

"Mom, can you slow down?" Carla gasped.

I did slow down, and then I stopped. Just ahead of us was a clearing where the kids had claimed the shelter. A father and three little boys were setting up a tent, and our mystery hiker was swaying in a net hammock tied to some trees.

I nodded hello to the others but turned to our boys with desperate eyes. "Where's the pump?"

"Way down the hill." Joel pointed.

Using a rickety privy, Carla and I tugged on our swimsuits. It was too civilized here to wear less. And we scrambled down the hill. While Carla pumped, I stuck my head under the ice-cold spray, gasping and giggling, rinsing off the day's sweat.

Carla hesitated then took the plunge, screaming in shivers the whole time she had her head under the water stream. With a washcloth, we scrubbed the rest of our bodies. After rinsing out a sticky, syrup-coated food bag, we used the bag as a washing machine for our clothes. We could squish and roll the bag, getting ultra-modern agitation action. After drinking our fill and refilling our canteens, we started back up, refreshed.

Vignette 100: The Green Monster
Friday, June 22, Day 16

Set a guard, O Lord, over my mouth; keep watch over the door of my lips

(Psalm 141:3 ESV)!

Tom had arrived at the camp and was talking to the mystery hiker. "This is Tara."

Tara smiled from her hammock. Tom frowned at me. "You should have told me you were going ahead. I was quite worried that you made a wrong turn."

"Sorry. I just kept picking up the pace, excited to wash up."

Tom nodded in his blank way, like when he doesn't hear what I'm saying anyway. My remorse was short-lived.

"Worried-schmorried." I mumbled as I stomped across the camp and set up my kitchen.

Did Tom help around the camp, like a concerned husband? No. Did he even look in our direction, remembering that he had a family? No. He continued in conversation with Tara on the other side of the clearing, too far away for me to hear even snatches. I slammed down our one plastic plate.

The boys had begun the long process of cooking rice pudding earlier, bringing it to a boil and then letting it sit for thirty minutes. I just needed to bring it back to a boil for a short simmer. I also had to string a line to hang our laundry. Fuming, I glanced at Tom and that girl, Tara.

While the rice finished simmering, I counted out an equal number of crackers for each of us, looking for a place to put Tom's pile. I glanced in his direction again. Big, black ants scurried everywhere. I hesitated, searching for a place to put Tom's crackers, thinking a few ants wouldn't hurt them. Finally, I found an ant-free rock for Tom's food. I hoped, sort of, that his would stay ant free.

Next, I spooned equal amounts of spam onto each of our crackers, one at a time. The kids stood around me in an eager semi-circle, waiting for each spoonful. Mama bird feeding little birds. Where is papa bird? I left enough spam for Tom in the bottom of the can and dumped his crackers on top.

Popping Popcorn

Exasperated, I looked at Tom again. "Tom." I tried to sound sweet. "It's time to eat."

He interrupted his conversation with Tara and squinted in our direction with little recognition. "Oh, yeah," he said, coming back into focus. "I'll be there as soon as I wash up."

I ignored him, busying myself with serving the rice pudding, while he grabbed his towel and toiletries and headed to the pump. I was over-the-top irritated when he sauntered into camp forty-five minutes later. We'd already done the dishes, and the boys had popped a couple of batches of popcorn over the campfire.

"That felt wonderful," Tom said. "I washed my hair, my body, my clothes, and had an interesting conversation with a park ranger."

I gave Tom the cold shoulder. Another interesting conversation, I snorted to myself. Oh, and how wonderful that you took the time to wash your whole body. How could he be so selfish to only please himself all evening while I was in charge of keeping everyone happy? Tom ignored my ignoring which fueled the flame. My thoughts were not God-like.

Vignette 101: Real Religion
Friday, June 22, Day 16

The Lord is compassionate and gracious, slow to anger, abounding in mercy. He will not always accuse. He will not keep his anger forever. He does not treat us as our sins deserve. He does not repay us according to our guilty deeds

(Psalm 103:8-10 EHV).

I continued to ignore Tom as we shook out our sleeping bags and lined them up in the three-sided shelter. Tom's was against one wall, mine was next, then Carla, Joel, Caleb, and Ben.

As we scooted into our bags, I flopped over on my side, giving Tom the cold shoulder. I sighed dramatically.

Tom touched my shoulder. I shook off his hand.

"What's the matter with you?" he whispered.

"Nothing," I hissed.

"C'mon. There's something going on."

"Really, you can't figure it out? You flirt with a cute young woman for an hour, then you take a leisurely bath for another hour, and you saunter into camp after I've done everything. How selfish can you be?"

Silence.

"Don't you have anything to say for yourself?" I baited.

"I'm sorry," Tom said. "I should have helped."

"Yes, you should have."

"Tara is a nineteen-year-old girl who has a summer job with the YMCA, leading hikers. She's getting paid to scout out this section of the trail and will be done at High Point State Park."

My interest was piqued. "Go on."

"We got into a discussion of her philosophy of life which led into a deeper discussion. She went to a Catholic high school, but turned away from God. She has searched out other religions but has found nothing that appeals to her. She believes that if she is good enough, God will accept her."

Tom stopped and sighed. "I told her she has salvation only through faith in Christ and encouraged her to study the Bible. It was an important conversation."

My anger vaporized. While I was busy being Martha, Tom was being Mary, sharing his Savior with a lost soul. I've always thought that Martha wasn't completely wrong. She had a houseful of guests to take care of. And I don't believe that Mary was completely right. She could have balanced her life a little better. They complemented each other.

This encounter was an example of how Tom and I often complement each other too. His strong points are my weak ones and vice versa. It doesn't always make for a smooth walk through life.

I rolled over. "I forgive you."

"I forgive you too."

Vignette 102: Pesky Pests
Friday, June 22, Day 16

You will not fear . . . the pestilence that stalks in the darkness

(Psalm 91:5, 6 ESV).

I wasn't riled up anymore, but I couldn't sleep either. "Are you still awake, Tom?"

"Yes." He inhaled deeply and blew it out.

"Tomorrow, let's dry out at High Point State Park. I saw in the guidebook that there's a beach and a bathhouse. Wouldn't it be dreamy to sprawl on the warm sand and let our damp clothes dry on rocks or bushes?"

"Sure, let's do that."

I smiled to myself, knowing I could probably ask for anything right now and Tom's lingering guilt would prompt him to say yes. But that's all I wanted—a few hours on a beach to dry out.

"Are you two done chatting over there?" Ben whispered through the darkness. "I can't sleep here, stuck against the wall. I feel like ants are crawling on me. You wanna trade, Mom?"

"Well, that sounds appealing, Ben." I shuddered.

"You're fine, Ben," Tom said. "Ants sleep at night."

"It must be my imagination." Ben rolled over.

"Do ants sleep at night?" I whispered.

"I don't know," Tom whispered back. I snickered.

A few black ants had been inside the shelter. We had brushed them out before we climbed uneasily into our sleeping bags.

I brushed something crawling across my neck and pulled my sleeping bag over my head. "I don't think they sleep at night."

Tom smacked his head and brushed at his hair. "I think you're right."

"How are we going to sleep?"

"We probably won't."

There was more activity than our own squirming—the ants were marching two by two or three by three or ten by ten.

~Distance hiked on day 16: 16.5 miles
~Brink Road Shelter to Mashipacong Shelter

Vignette 103: Going Downhill
Saturday, June 23, Day 17

Why are you so depressed, O my soul
(Psalm 42:5 EHV)?

After a scant night of sleep, I stumbled through breakfast prep, heating water for hot chocolate and then stirring Malt-o-Meal into the remaining hot water. Since Tara had not roused from her tent, we whispered as we ate and tiptoed from camp at 8 a.m.

The boys were out of sight before we reached the pump at the bottom of the hill. I wished I had their energy. I wished I had slept through the ant invasion during the night. And I wished this hike was over.

The clouds thickened, shrouding us and the trail in fog. From the mist leaped a deer. It bounded across the trail and disappeared silently into the mist again. Normally, I would have exclaimed and pointed at the deer. Instead, I sighed, wishing this hike was over.

For awhile, Carla and I walked ahead of Tom. She taught me a song.

Sing your way home
At the close of the day.
Sing your way home;
Drive the shadows away.
Smile every mile,
For wherever you roam,
It will brighten your road.
It will lighten your load,
If you sing your way home.

The song gave us good advice, to sing our way home, but I wanted to be home. I didn't want to sing. I wanted this hike to be over.

Mid-morning, Carla and I stopped at a rock outcropping where the views were spectacular on a clear day. Since the clouds were getting darker and the fog had turned to mist, no views were visible. I sighed again. Tom caught up with us, and we sat in the mist, staring at the non-view, snacking on pineapple rings.

We'd barely resumed our walking when the mist became a rain, forcing us to don our rain gear. My raincoat smelled like an old dishrag. The sleeves were still damp from the last storm. As soon as the rain slowed, I yanked off my raincoat and draped it over my pack, deciding I'd rather get wet from the refreshing rain than from a sweaty raincoat. Actually, I didn't want to get wet at all. I wanted to be dry at home.

Carla caught a newt again and wanted it to be her pet again. For an hour she walked, cradling it between her hands, distracted and tripping often. Finally I convinced her to free the newt so she'd be able to catch herself if she slipped on the wet rocks. She did.

Five minutes later, she slipped on a wet rock, fell, and skinned her knees. Minor first aid needed. I blew out a big breath. When would this hike be over?

Vignette 104: Sinking Lower
Saturday, June 23, Day 17

O my soul, why so disturbed within me
(Psalm 42:5 EHV)?

My motherly instinct squelched my self-pity. I hadn't seen the boys for over two hours, and I was anxious to get to the ranger station at High Point State Park.

"Let's hurry," I said to Carla, but we had barely kicked into high gear when we emerged from the trail onto the mowed grass. Nearby were a parking lot, a pond, and a big-log ranger station. It was 11:15 a.m.

The boys were scrambling around the pond, Caleb with fishing pole in hand.

"Look at this ten-inch bass," Caleb called as we approached, then he tossed it in. "A ranger said we could fish here all we wanted, but we had to throw them back."

"What took you so long?" Ben grumbled. "We've been here since 9:45 a.m."

"Two of our packs broke." Joel looked happier than he should. "I guess we'll have to quit now."

"Now there's a thought," I whispered. A good thought. Joel's suggestion broadened the crack in my resistance.

We were at the park where we planned to swim, shower, relax on the warm sand, and dry out our clothes, but it was cold and drizzling again.

"Let's take a look at the broken packs." Tom acted extra enthusiastic to make up for the gloom hanging over me.

"Minor repairs," he announced. Joel's needed a couple of stitches, and Ben's, a paper clip to replace a lost ring. Tom took over the task.

I slouched on a bench near the pond and got out lunch which was nut mix and another granola bar. It only took seconds for me to devour my share. I was still hungry, plus tired and depressed. The sky continued to threaten rain. I sat. The kids were squealing and jumping, catching fish and finding snakes.

A pay phone booth was in the parking lot. "Hey, Tom." I pointed to the phone. "Maybe we could make arrangements to get our car."

"Not yet." He was still in his extra-enthusiastic mode. "But I will call the monastery. Remember, that's our new goal for ending the hike."

He returned in a few minutes. "Yes, the monastery will be open to hikers next week when we plan to arrive. Aren't you excited?"

It seemed an eternity away.

Tara hiked in and set her pack on the bench next to me. My animosity of the evening before was forgotten as I gave her a friendly greeting.

"Whew, it's good to be done." She collapsed onto the bench. "I'd planned to loaf here today, but since it's so gloomy, I think I'll call up my boss to pick me up. A shower at home will feel great."

She trotted off to use the phone. My depression deepened. I longed to be done. I craved a shower at home. Before long, Tara was back. Soon she had a little pot of soup simmering for her lunch. My mouth watered, and I glared at my bag of nuts. Even M&M's couldn't cheer me.

Vignette 105: Really Low
Saturday, June 23, Day 17

Why so disturbed within me

(Psalm 42:5 EHV)?

Since the kids were entertaining themselves, Tom and I strolled into the visitors' center, leaned our packs against the wall, and sank into a padded, leather couch to write in our journals. To most people on a muggy, summer day, air conditioning would be a relief, but not to us in our sweat- and rain-soaked clothes. We shivered. Tom coaxed his arms into his raincoat to warm up and continued writing. I shuddered at the thought of sliding my arms into my sticky, smelly raincoat.

"I'm going to see what's happening around here." I jumped up, swinging my arms to warm up. I headed across the room, approaching the receptionist, Karen, according to her name badge.

"Hi, Karen. Can you tell me about nearby hotels, laundromats, and fast food. Especially fast food?"

She whipped out a map spanning twenty miles and pointed out the attractions. As Karen droned on, giving her canned talk prepared for the typical tourists, my focus faded. We were not your typical tourists.

"At the bottom of the hill is Port Jervis," Karen was saying. "You'll find a McDonald's and a DQ there."

I snapped to attention. "How far is that?"

Karen perked up. "Oh, it's down the hill, not far at all, probably five or six miles."

I shook my head and leaned onto the counter.

Karen glanced around, worried about having a dissatisfied customer. "Really, it's only a ten-minute drive."

"Thanks anyway."

Five or six miles downhill meant five or six miles hiking back up hill. I plopped back down by Tom to think.

Why are you downcast? Would it be worth a day's roundtrip hike to eat at McDonald's? Would we want to hike for twelve miles beside a busy highway with our four kids?

O my soul? But my mouth was yearning to bite into a Big Mac. My body was crying for warmth and dryness. High Point was my low point.

Vignette 106: Brief Reprieve
Saturday, June 23, Day 17

Bless the Lord, O my soul, and forget not all his benefits, who redeems your life from the pit, who crowns you with steadfast love and mercy

(Psalm 103:2, 4 EHV).

After a few minutes, I bolted upright in my chair, remembering Carla's bite, chastising myself for my negligence. The bite had been changing, still swollen but now covered with a rash and red scabbing.

I nudged Tom on my way to the door to call in Carla. "Let's have a ranger look at Carla's bite."

In a few minutes, Karen, the receptionist, studied Carla's bite. "This doesn't look good. I think it's a deer tick bite." She motioned to the ranger standing nearby.

Ranger Ted sauntered over, slipped on his glasses, and studied the bite, then frowned at me. "If she were my child, I'd have a doctor look at it. We've had several cases of Lyme disease around here."

I knew that Lyme disease was dangerous. I'd studied the symptoms and the disease before we left home. I also knew it couldn't be detected at this early stage, but the ranger's words *if she were my child* kept bouncing around in my head.

I hesitated, thinking what to do.

"I'll drive you into the hospital in Port Jervis," Ranger Ted volunteered.

I glanced at Tom. He nodded. It was decided.

"I'll be outside by the car." Ranger Ted headed out.

As Tom, Carla, and I hurried to our packs to regroup, Karen called after us. "Ask for Dr. Dille. She has studied Lyme disease extensively."

"We can't all pile into Ranger Ted's car." Tom glanced at me. "You take Carla, and I'll stay here with the boys."

"Okay." I dug into my pack for my wallet. "See if Carla and I may leave our backpacks here until we return." I tossed my sweat-stained hat onto my pack.

"Check!" Tom led us out the door to Ranger Ted's car.

"Hey, boys, come over here," I called.

They trotted over. Concern lined their faces when they saw Ranger Ted opening the back door for Carla and me. Ben broke the tension. "What's going on? Are Mom and Carla getting arrested?"

I raised an eyebrow and all giggles stopped. "We're getting Carla's bite checked."

"Good idea." Caleb gave Carla a pat on the back.

As the boys gathered around Ranger Ted's car to wave good-bye, the sun peeked through the thinning clouds. I paused, absorbing the momentary warmth, wishing for my promised nap on the beach.

Tom shrugged and hugged Carla, then me. "Sorry, but this is the best scenario. I'll take the boys to the beach." He leaned closer to my ear. "Bring us some real food."

Carla and I climbed into the back seat. Ranger Ted leaned in. "Please buckle up. I want to make sure you're safe."

I sank into the seat, feeling safe and pampered.

Vignette 107: Heading Lower Again
Saturday, June 23, Day 17

Yet I am poor and needy. May the Lord think about me. You are my help and my deliverer

(Psalm 40:17 EHV).

Ranger Ted zoomed downhill the six miles into Port Jervis, New York. At the outskirts of town, I spotted the DQ and McDonalds, remembering Tom's last words.

In a few minutes, Ranger Ted stopped in front of the hospital and put his elbow on the back of the car seat. "Here we are. Call me when you're finished."

The ER was filled with cases: a family praying for a dad with heart problems, a little boy crying with poison ivy, a teen cradling a broken arm, and more. We registered, requested Dr. Dille to an unfriendly receptionist, and settled into two seats in the far corner.

Only then did I notice the stares. I crossed my legs and peered out the window, remembering how awful we must look. Once Ranger Ted had offered us a ride, we didn't take time to clean up. My oily hair was flattened from several days of wearing my floppy, fishing hat, which I had, at least, politely left with my pack. I wished I had it to pull over my face.

I glanced down. Mud was caked under my fingernails and coated my legs, plus little piles of dried dirt had chipped off the tread of my hiking boots onto the sanitized, highly-polished floor. Mud-streaked Carla was oblivious to everything but the corner TV.

"Carla," I whispered, "Come with me to the bathroom."

There we scrubbed off the dirt but couldn't make the scratches and bruises disappear. We combed our hair with our fingers, glared at our clothes with disgust, and returned to more stares in the waiting room.

Those people thought we were homeless. Should I stand up and announce that I have a college education and that we're hiking the AT for a wholesome experience with our children? Should I announce that I have a closet full of nice clothes at home and that I shower daily?

Wisely, I decided against more of a scene than we'd already made. Being scorned was a new and uncomfortable feeling. I felt inferior, my inferiority reflecting in the gawkers' eyes. Who cared what they thought? I slouched a little lower in my chair, still caring.

Vignette 108: How Low Can You Go?
Saturday, June 23, Day 17

Give justice to the weak and the fatherless; maintain the right of the afflicted and the destitute. Rescue the weak and the needy

(Psalm 82:3-4 ESV).

An hour later, we followed a white-garbed nurse down the shiny hall. I willed the dirt to stay stuck in my boot treads. It didn't. A faint trail of dried mud and pebbles followed us.

"May we see Dr. Dille?" I tried to appear poised.

"I'm sorry. She's not on duty today." The nurse clipped her words.

"May we call her?" I felt desperate, knowing that our reason for suffering this humiliation was slipping away.

"If the emergency room doctor needs a second opinion, then he may call her." The nurse left, slamming the door.

I regretted following my motherly instincts, allowing guilt to control my brain. I knew we were wasting our time and money, but I felt trapped into following procedure. Before long a greasy-haired—at least we had something in common— doctor strolled into the room.

"And what is your problem?" he asked Carla in a thick accent.

She pulled back, tongue tied.

"We'd like you to check a bite," I intervened. Carla pulled down the neck of her T-shirt for inspection and stared at the doctor with worried eyes.

"Do you think it's a deer tick bite?" I asked.

He glanced at the bite a second, only a second, and then said, with exaggerated patience, "It's too early to tell. I will prescribe some ointment to relieve the itching and then you watch for these symptoms…"

I knew the symptoms by heart. I'd read and reread them a hundred times before we'd started this hike. I was tired of being moved along like a lame-brained number. I didn't want his ointment, and I still wanted to see Dr. Dille.

When the nurse returned with the prescription, I made one final attempt, "Now may we see Dr. Dille?" A steely stare gave me my answer.

I folded the prescription and shoved it into my pocket with no plans of filling it. We had plenty of anti-itch ointment in our bags.

There was another obstacle. Since the hospital was in New York and High Point was in New Jersey, I couldn't just drop a quarter into the pay phone and dial. Remember, we were in the pre-cell-phone era. I didn't have the area code for the New Jersey State Park, and the information desk didn't either. Finally, a kind-faced volunteer, seeing my frustration, found the number. Next I had to decipher the calling card information on our card and eventually got connected.

Vignette 109: Looking Up
Saturday, June 23, Day 17

The poor will eat and be satisfied. Those who seek him will praise the Lord
(Psalm 22:26 EHV)!

"It will be a little while," Ranger Ted said from his end of the phone line.

"Take your time." I smiled into my phone. Such a kind man. "We'll wait outside."

I had to get outside, away from that institution. Carla and I sat on the soft, green grass under a now-sunny sky. My frustration melted in the sun's warmth. Before long, Ranger Ted pulled up.

"How did it go?" He jumped out of the car and held the door for us.

With careful tact, I explained our hospital visit, trying to not belittle his plan and his assistance. But when I mentioned that I wasn't bothering with the prescription, a look of knowing empathy flickered across his face.

I had one more pressing request. "Would you mind stopping at McDonald's so that I can buy some real American food for my family?"

I knew he was on duty and had already done more for us than I could have asked. I held my breath.

"Oh, sure," he said, grinning.

With our supper packed in three paper bags and our retrieved packs beside us in the car, Ranger Ted drove Carla and me to the beach. We smothered him with thanks as he helped us unload, and we stood there waving until he turned the corner.

"Look, there's Joel, running to the beach." Carla pointed. "And the others are already by the water."

"Joel," I called. "Go get everyone. I have supper." With an exultant smile, I held up three McDonald's bags.

Joel's mouth dropped in amazement. Then he turned and raced down to the beach yelling, "McDonald's is here."

Carla and I spread our feast of Big Macs, french fries, and chocolate shakes on a big rock, unconsciously ignoring the many picnic tables nearby.

Caleb was the first to arrive. "Joel said that McDonald's is here." Caleb's whole body quivered with joy when I handed him a Big Mac, fries and a shake. Ben and Joel bounded in right behind him with outstretched arms.

"Real food!" Tom said moments later, just as excited. "You remembered."

We exchanged smug grins. He knew I considered all the trail food I had packed as real food, but today I had to agree with Tom.

"We thank you, Lord, for this food."

Sinking my teeth into that Big Mac was a pleasure! We were ravenous. The only noise escaping anyone wolfing down the food was an "Umm!" or an "Ahh!"

Vignette 110: Rest of the Story
Saturday, June 23, Day 17

Wait for the Lord; be strong, and let your heart take courage; wait for the Lord

(Psalm 27:14 ESV)!

When Karen came looking for us, our delicious, cholesterol-filled meal was finished. She held out a pamphlet, opened to a photograph.

"Here, I wanted to show you these pictures of the deer tick bite and rash. I think it looks just like your little girl's."

We crowded around, studying the picture of a deer tick bite. I was torn between comforting Carla and not offending this concerned lady. Why was my heart again taking over rational thought? "I'm not sure it looks exactly the same," I said, lamely.

Karen stared at the pamphlet. "And did you know that Dr. Dille was in her office, right down the hall from the ER? I wish I had just sent you to her."

So did I.

I shook my head. "Why wouldn't they let us see her? It was hospital protocol over patients. But thank you for all your concern."

Karen nodded, defeated, and strolled back to the ranger station.

Later, while Tom and I lounged on the beach, I pondered Carla's bite. "Do you think we should call Dr. Dille?" I voiced my concerns.

"I was thinking the same thing."

We walked the half mile to the pay phone. For forty-five minutes, we tried Dr. Dille's busy phone.

"Okay," Tom said. "I guess we just have to put Carla's bite in God's hands and out of our minds. She's fine, and we can't do anything about it now anyway."

"And when we get home, we'll all be tested for Lyme Disease," I added. "That's the earliest a doctor will put us on antibiotics anyway."

The rest of the story: We think our state park friends did one more good deed for us. We never got a bill from the hospital. Maybe patient concern did outweigh protocol. The best news? Six weeks later, we all were tested for Lyme disease. Everyone tested negative, including Carla.

Vignette 111: Finishing the Day
Saturday, June 23, Day 17

You are a hiding place for me; you preserve me from trouble; you surround me with shouts of deliverance

(Psalm 32:7 ESV).

It was late afternoon when we set about doing the longed-for, relaxing chores of the beach, like drying socks on rocks and hanging damp shirts on benches. While Tom and the boys took a dip in the lake, Carla and I showered in the bathhouse.

The faucets spit out such icy pellets that I think they were glacier fed. Carla and I screamed and giggled as we darted in and out of the spray, soaping up then rinsing in spurts. We emerged triumphant, with sweat-matted hair now fluffy.

Back at the beach, the guys were ready. I swung my pack on, the pack I hated two weeks ago. Now I did not feel completely dressed without it. As we tightened straps and snapped buckles, a few punks lazed on picnic tables nearby, smoking and watching us through slit eyes.

"Where ya headed?" one slurred.

"We plan to hike across the Hudson River before we quit." I shrugged. Their eyes opened wide in amazement. I couldn't resist adding, "We've already hiked 200 miles." Their mouths dropped open as we tightened our hip belts, like gunslingers' holsters, and hiked off into the sunset.

Nearby the High Point Monument, resembling the Washington Monument, towered above us. The boys and Tom had climbed it.

"It's closed now." Tom grabbed my hand and tugged me up the steps. "But come see the views from the bottom platform."

"We can see for miles and miles." I pointed into the valley. "There's Port Jervis."

"Over there are the Pocono Mountains." Tom motioned in the direction we were heading. Fertile fields crisscrossed the valley in between.

"Mom, did you know that on a clear day New York City is visible?" Joel asked. We strained to see, but the evening haze screened it from our eyes.

Dusk was rolling in when we left the monument for the short walk to the three-sided shelter. It was teaming with life. Nestled inside, peering out from their sleeping bags, were several new faces. They introduced themselves as Poly Esther, The Archer, and One Braid. Another face, this one familiar, belonged to Foxy, the hiker we'd met at Delaware Water Gap, the suspicious young man who turned into a toilet paper professional. We chatted briefly with the group, but darkness was creeping in. Across the grassy patch, the same father and his three sons were camping. "There's a rock-free spot for your tents over here," he called.

While the guys set up our tents, I attempted to make a fire. No success. The boys nudged me aside but were unsuccessful too. Finally, Tom strategically planted kindling and soon had a roaring fire. There we had a family devotion and, once more exhausted, climbed into our tents.

I felt safe and snug in our tents, surrounded by family and friends. I was again content with where I was at this moment. As I drifted off, the tune of *Sing Your Way Home* danced through my mind. Thank you, God, for bringing me back home, right here.

~Distance hiked on day 17: 7.2 miles
~Mashipacong Shelter to High Point Shelter

Vignette 112: Detestable Manna
Sunday, June 24 – Day 18

He rained down on them manna to eat and gave them the grain
of heaven
(Psalm 78:24 ESV).

"It's only 6:30." Tom tugged on the food bag rope and eased it to the ground. "Do you think we can get an early start?"

I pulled out my daily list. "Blueberry oatmeal pancakes made with amaranth flour and ground almonds are on the menu. They're super nutritious, but they take longer." I turned to Tom. "Don't we want to have a nutritious start to our day?"

Tom nodded, reluctantly. "Sounds like God's perfect manna."

"And I have blueberries for an antioxidant, but there is a problem." I frowned. "I didn't soak the dried blueberries over night, but don't worry. I have a plan."

I had several good reasons for not soaking the berries. We got in late. I didn't have a container in which to soak them. I could have soaked them in a bag, but where would I have put the bag overnight so it would be safe from critters? My best excuse was that I was tired.

Since our water source was a creek, and I'd have to boil the water to purify it, I decided to boil the berries and then use the water for my liquid. The plan could have worked, perhaps, if I'd allowed the water to cool or dipped out the berries, but I didn't. Time was flying. I couldn't wait. I dumped the hot water and blueberries into my oatmeal/whole grain pancake mix and stirred frantically. The batter coagulated into a thick paste. The batter looked like mush before I fried it and the pancakes looked like mush after they were fried. Still, they were nutritious.

As usual, I cooked one big pancake—I mean mush cake—at a time on our one-burner stove, but with the canister almost empty, the cooking went even slower than normal. Hadn't we remedied this problem two weeks ago?

"Time for mush," I called to the next person, and he or she would scurry over. Fifteen minutes later, the next mush cake would be ready and another call would be issued. Finally, triumphantly, I handed out the last mush cake.

"Are there seconds?" Ben hovered nearby.

"Seconds?" I sputtered. "I've already been cooking for one and a half hours." I glared at Tom with stormy eyes. "This is it. I'm not making pancakes again."

It was the desperate statement of a woman who had lovingly assembled each pancake mix at home, being sure they were highly-nutritious, total-protein meals.

"I don't care how nutritious they are," I continued my ranting. "This is it. I detest this food. From now on, we will eat simple, packaged food."

"We could always eat more granola bars." Tom sighed.

Vignette 113: Playtime
Sunday, June 24, Day 18

Then our mouth was filled with laughter, and our tongue with shouts of joy

(Psalm 126:2 ESV).

While I cooked one mush cake after another, all was not misery. The brook was babbling down the hill, and the sun was making a patchwork quilt on my rock-table.

The best thing was that the kids were absorbed in another made-up game. The object of the game was to walk the length of a bouncy, two-inch diameter limb. It was wedged at both ends into rocks and formed a balance beam a couple feet off the ground.

The sure-footed boys tried first. No problem. Carla was next.

"Watch this." Carla paused at one end, arms lifted Olympic-style, and pranced across.

"Ah, that's too easy," Caleb said. "Let's see if we can shake you off."

The competition amidst giggles continued.

While Ben was waiting his turn, he bent a two-inch sapling and watched it spring up. "Do you think it would spring up if I held on?"

"No, you're too big," Joel said, "but it might work with a smaller body, like Carla's."

All three boys turned to Carla. "Wanna try?" Ben asked.

Carla grinned. Of course, she did. And so the boys launched Carla just to swing over their heads and down the opposite side. More giggles.

Ben was playing along with the younger kids, having a great time. Back home he'd be too old to have fun like this. He'd have to act condescending to his siblings and indifferent to fun.

Tom finished the packing-up chores and squatted beside me to watch the fun. We laughed along with the kids.

"I'm glad Ben gets this chance to be a kid," I said, then added, saddened by his vanishing youth, "This is probably the last time he'll really play." Little did we know he'd always love goofing around and playing games.

Vignette 114: Wandering in the Wilderness
Sunday, June 24, Day 18

Some wandered in desert wastes, finding no way to a city

(Psalm 107:4 ESV).

It was 9 a.m. when we left camp. We were apprehensive about the trail ahead. One book suggested catching a bus and just forgetting this portion. Another said it was a varied, surprising day of hiking. Our trail master, Tom, said we had to hike to the pavilion in Vernon since there were no campsites, no shelters, and limited water until then. That was seventeen miles away.

But fortified with our healthy mush, we accepted the challenge, and the boys took off ahead.

Stone fences were our first unusual encounter. We stumbled through gaps in about twenty of them. The trail was not as rocky as usual. All the stones were in the fences. One formed the New Jersey-New York border. We zigzagged into New York and back to New Jersey several times. The other walls were a mystery to us.

Switchback descents brought us gently off the ridge and into farmland. We skirted corn and wheat fields, ducked under barbed wire and electric fences, dodged cow pies in a pasture, and balanced across logs in a bog while cows breathed down our necks. All the time, we were trying to follow the white blaze marks which were supposed to agree with the intricate directions of the guidebook. Many times, Tom, Carla, and I retraced our steps, reread the directions and rescanned the area for those blazes.

The guidebook's not-so-easy-to-follow directions had us skirting fields, climbing questionable stiles, passing a hidden pond, crossing overgrown stone walls, and following a tree line that was actually scrubby bushes. Just when we thought we were hopelessly lost, a white blaze would appear.

As we struggled to coordinate our directions, we wondered if the boys were floundering somewhere in a wheat field on the wrong side of the upper pasture, but when we emerged from the lower pasture, the boys were visible sitting on a real stile. I recognized it. It was made with heavy wood and formed a staircase up one side of the fence and down the other.

Joel was tapping his foot, his turn to ask, "What took you so long?"

I ignored him. "How did you guys manage to find your way without a guidebook?"

"No problem, Mom. We just followed the white blazes."

"Well, I insist that you take the map, just in case."

As they hiked off, a little disgruntled, I winked at Tom. "Perhaps, reading the map and wondering if they've made a wrong turn will slow them down a little."

Vignette 115: Firm Footing
Sunday, June 24, Day 18

But as for me, my feet almost slipped out from under me. I almost lost my footing

(Psalm 73:2 EHV).

The AT has as many different and picturesque trail bridges as it has groups maintaining them. We often stood in awe to see the amount of work that hiking groups put into each bridge, not to mention shelters and their sections of the trail. It really is a sacrifice of love for the thousands of volunteers who labor to keep a more-than-2,000-mile corridor open.

But a well-constructed bridge was not the case at the next creek. The babbling creek was only about four feet wide and six inches deep. Really, why would you need a bridge? Instead, a sturdy log connected the two banks.

I studied the slight arch of the log, the curved top, the damp bark. My confidence in my balancing abilities melted away and ran down stream.

"Here, take this." I handed Tom the guidebook that I was carrying in one hand and took one careful step up. I shuffled toward the middle of the log, feeling my pack sway to the right. I paused. My pack swayed left. Did I look at the swirling water or over-react to my shifting pack? I leaned right, inching forward. My pack swung right.

"I'm losing it." I bent my knees, dug in my toes, and raised my arms like a tightrope walker. Nothing helped as my pack swayed in bigger loops

Rather than try to continue my balancing act, fail, and fall pack- or head-first into the water, I jumped off the log, hitting the water with a big splash, planting both feet in the middle of the stream. With spring reaction, I was out and up the opposite bank. Water had barely had time to penetrate my boots.

"Well done!" Tom and Carla applauded.

"I give you a ten," Carla added.

I poised with my hands on my hips. "Let's see you do it."

They, too, had trouble crossing the log, not because they lacked confidence, but because they were doubled over laughing.

Of course, the boys were waiting for lunch at the next road. They lounged on a step-like dirt-and-rock rise above the pavement. Litter from the ditch and an occasional junky car rattling by complemented our lovely lunch of almonds, dried bananas, and of course, granola bars.

Vignette 116: How Far Is It?
Sunday, June 24, Day 18

O Lord, all my longing is before you; my sighing is not hidden from you

(Psalm 38:9 ESV).

Only a quarter of a mile later, the trail crossed a dam at the end of a pond. It was shaded by trees, isolated from the world, and teamed with schools of goldfish.

"Wouldn't this have been a beautiful place to have lunch?" We inhaled the beauty for only a minute. The seventeen-mile day loomed. The trail emerged from the wooded area and climbed a grassy hill. At the crest, I glanced back.

"Hey, look!" I pointed to a surprising view mentioned in the guidebook. "There's High Point Monument, already behind us by five miles."

"Oh, yeah, it looks like a flagpole on the horizon." Tom narrowed his eyes. "Isn't it amazing how far we can walk one step at a time?"

The hot afternoon sun beat down on us as the trail meandered through fields of grass as tall as my armpits and taller than Carla's head. Sometimes the meanderings took us back in the direction we'd just come. We were close to civilization, crossing many roads, cutting through backyards. Once we circled a dilapidated farmhouse where a helicopter with props whirling sat in the barnyard. We wondered if drug dealers were getting ready to take off. We hoped the boys were safely ahead.

We'd hiked nine-and-a-half miles since morning when we reached State Line Road. The trail followed the road for a half mile. We strained to see ahead, hoping to catch a glimpse of the boys. Where were they?

Since our canteens were empty and no water sources were listed for miles, out of desperation, we stopped at a house to ask if we could fill our canteens at the outside faucet. The elderly lady invited us inside to her kitchen. Her husband, who had lost a leg to cancer, eased his way into the kitchen too. They seemed eager to have company, so we chatted a little bit.

"How far is it to Vernon from here? We are hoping to make it to the pavilion there." Tom knew our guidebook said eight more miles.

"Oh, I don't really know. It's quite a ways." The man shook his head. "Anyway that you go, you'll have to cross the Pochuck Mountains."

We didn't want to hear more about how far and how hard our day's hike could be, so we said our thanks and scurried out.

"Did you see those oranges on the table?" Tom asked as we cleared their yard. "Wouldn't a juicy orange taste good?"

"And did you see the Oreos in the cookie jar?" Carla asked.

I sighed. My whole body sighed.

Vignette 117: How Far Is It Now?
Sunday, June 24, Day 18

My heart beats quickly. My strength leaves me. Even the light of my eyes is gone from me

(Psalm 38:10 EHV).

Before long we turned right, entering a sod farm. The white turn blazes were nearly hidden. As the trail circled the farm on a gravel road, the white blazes became non-existent. To our right flowed the Wallkill River, and to our left, spreading like a half-mile green velvet carpet, were wide strips of sod. I couldn't believe that the boys had found their way. Again we scanned the distant edges, and again we saw no boys. We checked the riverbanks, too, thinking they might be fishing. They weren't.

"Where did you say we'd meet up?" I glanced at Tom.

"I don't think I did."

We decided the best route would be to continue, assuming the boys were ahead. Even though the trail was easy hiking, our steps dragged as we skirted the wide, open area. Being able to see how far we had to walk made the one-and-a-half mile trek seem much longer.

"Wouldn't that soft, green grass be a great place to set a tent?" Tom mused. It did look inviting.

When we reached the far end of the sod farm and turned left, heading once more toward Pochuck Mountain, we caught a glimpse of red and blue across the field.

"Do you suppose that's the boys?" I asked wistfully.

"I can't tell." Carla shook her head.

"Me either."

We kept our eyes riveted on that splash of color, but the closer we drew, the more it looked like farm machinery.

"Wait." I slammed to a halt. "Somebody is moving there."

"Yeah, I see them," Tom said. "They're fishing in that pond."

And they were.

"What took you so long?" Caleb ran to us. "Do you think we'll make it to the pavilion today? Do you think there's a basketball court there? Do you think somebody will loan me a basketball if there is?"

I shrugged to his barrage of questions and handed out pastry pops. The boys wolfed theirs down and took off again for our next meeting point.

Tom shouted after them. "County Road 565, four miles ahead. Be there."

We three floundered along in their dust. The boys were out of sight before we reached the first switchback of Pochuck Mountain.

Vignette 118: Popsicle Interlude
Sunday, June 24, Day 18

God has pitched a tent for the sun in the heavens. It celebrates like a champion who has run his race. It sets out from one end of the heavens. It runs until it reaches the other end. There is nowhere to hide from its heat

(Psalm 19:4-6).

The climb up Pochuck Mountain was steep, and Tom stopped often, gasping for breath. Finally at 5 p.m., we reached what we thought was the summit. Staring back, we could again see a hazy needle glimmering on the horizon. It was our last glimpse of High Point Monument.

I pulled the guidebook from my pack and read, "At four-and-a-half miles into this section, reach summit of western ridge of Pochuck Mountain and begin to descend."

I pointed to the page. "This is where we are." Tom nodded.

"Oh, no," I continued reading. "At 4.8 miles, begin steeper ascent on rocky trail. At five miles, reach crest of middle ridge. At 5.6 miles, continue to ascend, and at 5.8 miles, reach the summit."

Tom wiped his forehead with his sleeve. "You mean we have 1.3 miles of tough climbing ahead of us?"

It was tough, but we were rewarded with occasional summit views and a rustic bridge over a rushing stream. At 6 p.m., we reached County 565 and found the boys lounging in the tall, roadside grass.

"The lady in that house gave us a popsicle." Ben pointed across the road. I wanted to run over and beg for a popsicle too. My hot, dry mouth couldn't imagine such cool comfort.

"And she said that we could ask them for shortcut directions to Vernon when you arrive."

Relief filled me. We three newcomers who had not been blessed with popsicles traipsed across the road and up the driveway. We were seeking directions. And popsicles.

Two men came out on the deck. One brought us popsicles. We reached out with shaky, grateful hands.

The icy liquid glided down our parched throats.

"Yum!" Carla closed her eyes.

"Mm!" I joined her.

Those popsicles were refreshing, a gift from heaven, but lasted about as long as a drop of water on the rich man's tongue in hell. After hiking more than fifteen miles already, our dehydrated, starved bodies were crying out for more, but our Midwestern manners took over, and we showered the kind man with thanks.

He told us about the AT's climb across one more mountain before we could go directly into Vernon. He also gave directions for a shortcut via highways. We glanced at the sun on its downward orbit and set out on the shortcut, hopeful that we had, at most, two easy miles to go.

Vignette 119: It Can't Be This Far
Sunday, June 24, Day 18

Hurry, let your compassion come to meet us, for we are very weak

(Psalm 79:8 EHV).

The pavement was hot as our burning feet pounded along it. The six of us formed a single file line snaking down the road, clinging to the edge of the pavement as traffic roared past.

The boys no longer had the zip to go ahead, plus we'd left the comfort of those white blazes when we left the trail. The man's sketchy directions were getting sketchier. We walked, walked, walked, corrected our direction once by asking further directions, and walked some more. We passed houses which appeared to be at the outskirts of a town, but no town appeared.

Finally, when we reached a T-intersection, the kids and I were stumped. Now which way should we turn? Nobody remembered. I had been leading and glanced back. Tom had fallen behind a hundred yards.

Just then, a flashy motorcycle pulled up with a middle-aged man, decked out in neon nylon, perched on top. I flagged him down. "We're trying to find the pavilion in Vernon. Are we headed in the right direction?"

"Oh, yeah." He revved his motor. "Take a left here. Go for a couple miles, then turn—"

"What?" I pounced on his words. "How far is it to the pavilion?"

"Can't be more than four miles." He shrugged one shoulder. My mind and body went numb. Out of the corner of my eye, I saw the kids' faces crumbling.

Holding on to my composure, I thanked the man. He rode off, oblivious to our despair. Nobody with a set of wheels and a full stomach can imagine how desperate we were becoming.

Tom approached as the motorcyclist roared off.

"I can't make it another four miles." A sob locked in my throat. "We haven't had much to eat all day, and we've walked seventeen miles. It's going to get dark soon. How can I cook supper? What are we going to do?"

"We'll stop at the next house that has a pickup truck in the driveway," Tom said. "I'll ask for a ride to the pavilion."

Vignette 120: Yogi Bearing
Sunday, June 24, Day 18

Let them give thanks to the Lord for his mercy and his wonderful deeds for all people, because he satisfies the desire of the thirsty, and he fills the desire of the hungry with good things

(Psalm 107:8, 9 EHV).

Jim Smith was sitting down to a steak supper when Tom knocked. He said he'd give us a ride if we didn't mind waiting until he was done with supper. We rejoiced and collapsed in his luscious grass to wait, so relieved to have a ride that none of us even envied his steak supper.

Later, as the scattered town of Vernon whizzed past my cab window, I thanked the Lord for Jim Smith's ride. Even if we'd have had the strength to hike it, darkness would have covered us before we'd have reached the pavilion. Jim even stopped at the police station, because our guidebook said we should register there. The police didn't seem to care.

Mentally, I planned supper. I'd begin cooking tuna helper immediately while we still had a little daylight. I would joyfully cook, standing at a table and not squatting over a rock. My soaring spirits sank when we reached the pavilion. It was teeming with a family reunion.

Half-heartedly, I scanned the area. I spotted a water spigot on a nearby building and a patch of grass next to it where I could squat over the stove to cook a meal that wouldn't satisfy anyone. We skirted the pavilion, not wanting to disturb the family, and set up camp.

Emerging from the edge of the pavilion was Poly Esther, the hiker we'd met last night. She beckoned to us. "I've been Yogi Bearing up a whole picnic of food for you."

"What?"

"The family invited us hikers to help ourselves to their leftovers, and even though they're starting to pack up now, you're still welcome. It's quite a feast."

Our camp preparations screeched to a halt while her words sank in. "We may eat with them?" I stammered.

"We already did." Poly Esther nodded and pointed toward two guys sitting against a nearby shelter. The Archer and One Braid waved.

"Let's go!" Ben whooped.

Fried chicken, ham, potato salad, green salads, Jello, cheese, bread, baked beans, brownies, cake, lemonade. We tried to sit downwind from the daily-bathed family, but they flocked to us, piling on more food. We obliged them by shoveling it all in, struggling to balance good manners with voracious appetites. Family members chatted with us, chuckled at our food capacity, and graciously pushed more in our direction. We didn't refuse.

At last filled, we voiced our appreciation to the departing family and, like dumpster-diving bears, waddled with our protruding stomachs back to our campsite.

Vignette 121: Who's Out There?
Sunday, June 24, Day 18

If you make the Most High your shelter, evil will not overtake you. Disaster will not come near your tent

(Psalm 91:9, 10 EHV).

Soon the pavilion was emptied of its daytime guests. But we still had a few chores. We all rinsed out our clothes, wrung them tight, and hung them on the lines that Tom had strung in the pavilion. It was such a civilized thing, to dry our clothes away from the ever-present nighttime dew.

While the kids turned in, content and exhausted, Tom and I walked to the Mobil station and called my parents. "We're still alive," I said, "and we are cutting our hike short."

"That's such a relief," Mom said and whispered to Dad. I could almost see him grinning and nodding in the background.

We also called our friends Greg and Lisa in Fond du Lac. Lisa had been shipping the boxes of food that I had pre-packed and labeled at home. Only Greg was home.

"Hi, Greg!" Tom took the receiver. "We're cutting our grandiose hiking plan in half. Could you tell Lisa to not send anymore boxes?"

"That's great. We'll be happy to have you home early, but I have something for you to tell your boys." Greg had been the boys' grade school basketball coach. "I know they think your hike is ruining their chances of playing basketball, especially in the roles of leadership they crave, but that's not true. Tell them that Coach Wood of UCLA said his best players play no basketball in the summer. He told his players to get away from the game for the summer."

We could hardly wait to pass that tidbit on to Caleb and Ben who used basketball as one major reason to whine about the AT on a daily basis.

Finally, lying flat on my back in our tent, I was just drifting into a state of relaxed bliss when I jolted awake. Someone was rummaging through our packs. "Who's out there?" I squeaked, striving for but failing to use a voice of authority.

Caleb burst into giggles at my quavering question and mimicked my voice. "Who's out there?" he asked and laughed again. "It's okay, Mom. I'm just getting a drink of water."

Still snickering, Caleb zipped himself into the tent. I could hear him snoring in seconds, but he didn't forget the evening experience. "Who's out there?" asked in a quavering voice, became a long-standing nighttime joke.

~Distance hiked on day 18: 17.2 miles or so

~High Point Shelter to Vernon Pavilion

Vignette 122: The Tortoise and the Hare
Monday, June 25, Day 19

O Lord, hear my voice! Let your ears be attentive to the voice of my pleas for mercy (Psalm 130:2 ESV)!

Grocery shopping, post office food pickup, and general catch-up chores filled our morning. After a heavy lunch of tuna helper and a whole pie, and after we had distributed the food shipment among our packs, finally at 1:15 p.m., we dragged ourselves along Highway 94, exiting Vernon.

Heavier packs, heavier bodies, a hot busy highway. We lifted limp hitchhiking thumbs, not really expecting any help in traveling two miles back to the AT. Who would pick up six people with six backpacks? No school bus, super-sized van, or Jim Smith in his big pickup came along to oblige us, so we tramped on.

We intersected the AT beside a cornfield, not at one end or the other, but in the middle. So we walked through the cornfield with the sharp leaves slapping at our faces, arms, and legs. Who would plant a cornfield on the AT? Who would walk through a cornfield to stay on the AT? I shook my head. "We would, but why?"

And as we cleared the field and began to struggle up Wawayanda Mountain, my quiet question, like the rhythm of my heart, pounded in my head. *Why?* Step. *Why?* Step. But onward we trudged with our boys zipping ahead.

It was here that we passed a troop of Boy Scouts. To my glee, we stayed ahead of them. A moment's victory. A tad of spring returned to my step.

But not to Tom's. Once more Tom's pace had slowed to a crawl. Carla and I would walk five minutes and wait ten minutes for him to catch up, and then repeat the walk-wait pattern. Tom was the tortoise. Carla and I were the hares. Each time I became more impatient. Part of my impatience was that the boys were stretched out far ahead of us. I became more disturbed the farther behind we fell.

I try to be patient. I really do. I might even appear to be a patient person, but I like to clip along. Whether driving, walking, biking, or whatever, I hurry. It's a huge defect in my personality, but most of the time I can keep my mouth shut. I hate it when my mouth reveals the truth.

I snapped. "I can't stand to walk this slowly. It's 3 p.m., and we've covered less than three miles of the AT. You can just dawdle. I can't imagine your walking any slower. Carla and I are catching the boys."

I took a deep breath. Carla's eyes opened big with concern. Tom shrugged.

In his exhaustion, Tom didn't really care what I said or thought. And I didn't care that he didn't care. I stomped off, tugging Carla along, and muttering, "How much longer?"

Vignette 123: Becoming a Tortoise
Monday, June 25, Day 19

Be still before the Lord and wait patiently for him; fret not yourself

(Psalm 37:7 ESV).

In a half hour as my hurried pace left me breathless, Carla and I met up with the boys at an unusual bridge.

"Look how this is constructed, Mom." Joel pointed to the deck of the bridge. "See how the logs are crisscrossed with twigs."

"Nice." I struggled for enthusiasm. We teetered and twittered, prancing across the tricky bridge.

As the afternoon waned, Carla and I managed to keep up, Carla easily and me gasping. The trail had lots of ups and downs. "A real killer," I had written in my journal. However, I could have been killing myself with the breakneck pace of keeping up.

Finally, done in, I snapped again.

"Carla, you stay with the boys." I shot a glance at each boy. "And you stay with Carla. Got it?"

No questions. Four quick nods.

"Okay, then, I'll wait for Dad. We'll see you at the camp."

While they tiptoed away, I shrugged off my pack, dropped it into the middle of the trail, and plopped myself down, leaning against my pack to wait.

A song played in my mind, "I'm in a hurry to get things done. I rush and rush until life's no fun. All I really have to do is live and die. I'm in a hurry, and I don't know why."

The air was fresh. The greens of the trees swayed above me. I wasn't organizing, herding, or tapping my foot impatiently. I inhaled deeply and closed my eyes. My huffiness evaporated with my sweat. Perhaps playing the tortoise is smarter than playing the hare.

Vignette 124: Avoiding Mad Mamas
Monday, June 25, Day 19

Let my prayer be counted as incense before you, and the lifting up of my hands as the evening sacrifice! Set a guard, O Lord, over my mouth; keep watch over the door of my lips

(Psalm 141:2, 3 ESV)!

Tom and I dragged ourselves into camp at 6:20 p.m. Our accommodations were impressive. Another garbage can greeted me. It was tipped over, so I righted it lovingly, wondering if perhaps bears had tipped it over. Camp was also graced by a cement picnic table and dilapidated, but, still useable, lawn chair. And nearby but out of sight, a short walk into the woods, was an outdoor toilet. No walls, just a seat. We'd left modesty someplace back near Port Clinton, so we were okay with that not-so-private privy.

The kids had been busy making Mama happy. Caleb looked up from the wood he had piled, ready for a fire, and smiled. "Joel and Ben went to the ranger station to find water. I gave Carla a job too."

"We thought you'd like some pistachio pudding for our appetizer." She grinned as she whipped the pudding in our bowl.

I glanced sideways at Tom. He winked.

Joel and Ben returned with every possible water container filled. They lined up the water on the table.

"We're cooking tonight." Ben lit the stove and set a kettle of water onto the burner. "You two just sit at the picnic table. Carla, bring them the appetizer."

As the kids prepared and served our dinner, Tom and I lounged at the table, nudging each other. It was a feast. First we shared two freeze-dried meals the boys had nabbed from the grab box in Delaware Water Gap—barbecue beef and crackers and spaghetti. Then we continued with Hearty Hungarian and Caleb's popcorn over the fire.

Later while we were finishing our camp chores, including Tom hanging the food bag, two hikers passed through. "Hang that extra high," one of them warned. "A mama bear and her two cubs scavenge for garbage here."

The other hiker nodded toward the kids. "Never get between a mama bear and her cubs. You'll see what a mad mama looks like."

"We don't need a mad mama here." Tom grinned at the kids.

No mad mamas appeared.

Caleb's campfire crackled and sang all evening. We sang a little, too, and Tom read the sermon which Jerry Krug had sent us. All around, it was a peaceful evening.

~Distance hiked on day 19: 7.8 miles

~Vernon, New Jersey Pavilion to Wawayanda State Park

Vignette 125: Another State
Tuesday, June 26, Day 20

He brings peace to your borders
(Psalm 147:14 EHV).

We awoke to a beautiful sunny day and breakfasted on oatmeal, raisins, and brown sugar. Next we redistributed the food supply from yesterday's pick-up. My pack, now less than thirty pounds, was more manageable. Could a heavy pack be the reason for the mad mama of yesterday?

The boys took off before 8 a.m. We three followed shortly. A few minutes later we came upon their packs, set on the trail. They had gone below to fill canteens. When they emerged with full canteens and big grins, they handed ours to us and tore off as we plodded behind.

At four-and-a-half miles into our hike, we came to the New Jersey-New York state line.

"Look at this." Tom pointed to a white paint line streaked across the trail. On one side the letters NJ were painted, on the other NY. "You girls line up. Let's take a photo."

I squatted in New Jersey, Carla in New York, and we held hands across the state line. "Can you believe we're entering our third state? Good-bye, New Jersey. Hello, New York."

In fact, New York City's airports were only a few miles away. The many jets soaring overhead testified to that. Two weeks ago when I saw a jet, I wished to be on it, flying anywhere away from the Appalachian Trail. Now, on this beautiful morning, I was glad to be here, moseying in the backwoods, away from the panicked pace of people.

I paused in our hike to read aloud from the AT guidebook, "Climb steeply, then very steeply."

"Very steeply must mean to use ladders," Tom said. I glanced up to see Tom, in an unusual moment of being ahead of us, dangling from a ladder. We followed Tom, scrambling up that ladder and several more.

New York gave us a great first impression with views of blue, inviting Greenwood Lake below, highlighting the flat rock lookouts. The three of us stopped at Prospect Rock to bask in the views.

An hour later we met up with the boys to eat lunch. The views were still spectacular with Greenwood Lake ever visible and looking more and more inviting as the hot sun bore down on us. Lunch was nut mix, fiber bars, and dried pineapples. We gagged down our lunch, weary of the monotony of nuts, granola bars, and dried fruit.

After lunch, the boys scampered ahead. Carla and I waited for Tom to reapply moleskin to his feet, Vaseline his side, and wrap his knee. Impatience poked me, but I did not cross the line.

Vignette 126: Is Roger Home?
Tuesday, June 26, Day 20

It is well with the man who deals generously and lends

(Psalm 112:5 ESV).

We'd only walked a short distance when we came to a trail register and a blue-blazed side trail heading down the ridge. The boys were lined up across the AT, arms crossed, blocking our way.

"There's a note here that says Roger's cottage is open." Ben pointed to the little square of paper tacked to the blue blaze.

"Can we check it out?" Caleb asked.

"He must have arrived early." Joel pointed down the side trail, eyebrows raised.

We'd read about Roger's cottage in the Philosopher's Guide. We'd memorized the details. Roger opened his cottage to a maximum of four hikers per night for several weeks each summer. He fed them, gave them a ride to the lake to swim, offered them a shower and a bed for the night, and breakfast the next morning. We figured we were too large of a group to barge in on Roger, but it would still be nice to meet him.

The Philosopher's Guide also said that Roger would not arrive until June 29 and, last night when we reviewed the next day's hike, we had glumly conceded that we'd miss meeting this benevolent, legendary Roger, a 2,000-miler himself.

But now with lighter steps, we slipped and skipped down the half-mile trail to Roger's cottage. Of course, the kids had raced ahead and were nowhere in sight when I approached the cottage. And Tom was quite a distance behind me, cushioning his knees with each step. I peered around hesitantly and then saw Carla's grinning face between the lace curtains, her nose pressed against the screen of the window. She pointed to a ladder, and I scrambled up to the deck.

Inside the little cottage, the kids were sitting on the couch, sipping sodas.

"Well, you must be the mama." A man with graying, wavy hair and warm smile handed me a cold ginger ale.

"We told Roger that you'd like ginger ale," Caleb said.

Roger's eyes crinkled. "Nice kids, Mama."

I shook Roger's hand. "I'm Janie, and I love ginger ale. Thank you!"

"As you can see, I just arrived at the cottage." Roger pointed to the dusty tables and carpet. "But what kind of soda would your husband like?"

"A Coke," all four kids shouted.

When Tom arrived in a few minutes, Roger handed him a Coke.

"We just wanted to meet you," I stammered. "Of course, we know that we are too many for you to host."

"Don't be silly." Roger shrugged one shoulder. "I'd love to host you tonight."

The kids' beaming faces reflected Tom's and my pleasure at the thought of staying here. A cottage with a roof. Windows with screens. Couches and chairs. A welcoming man who hands us sodas and even cooks. How could we show our appreciation? We helped Roger with a few chores like shaking out rugs, vacuuming the back porch, sweeping cobwebs from the doorframes, and more.

Vignette 127: Roger's Hospitality
Tuesday, June 26, Day 20

My soul will be satisfied with rich food. My mouth will praise you with lips filled with songs

(Psalm 63:5 EHV).

At 3:15, Roger gave us a ride the short mile to Greenwood Lake. But the sun disappeared behind a cloud, and the lake that looked inviting as we sweated along on the ridge above, seemed to turn its cold shoulder to us.

I perched on the beach, trying to absorb some warmth from the sand. Tom had water fights with the kids, splashing and throwing them around. Their giggles almost made my goosebumps worth the shivers. But I really wished I'd stayed at the cottage to take a hot shower and nap. When the water fun waned, Ben and Carla dropped down beside me.

"I'm freezing," Carla said.

"Should we head back to Roger's cottage?" I asked. "A hot shower will feel great."

"Let's go," Ben said. Carla jumped up.

The three of us hiked back up the hill, another short mile, while Tom waited with Caleb and Joel who were, of course, fishing.

Carla showered first. Soon I melted into the steamy shower, too, warming up, scrubbing up, and filling up with contentment. Later, refreshed, I popped into the kitchen where Roger was cutting, sizzling, and arranging food.

"How can I help you?" I asked.

"In a few minutes you may set the table on the deck, but for now, just relax. Enjoy some down time but be sure your family is ready for supper. It's served at 5:30 sharp."

At 5:30 p.m. sharp, we were all in our places with bright shining faces, seated around a real table on Roger's peaceful deck. Slivers of early evening sunlight shimmered through the trees. A soft breeze caressed our skin. I glanced at the faces of my family. Everyone was smiling at Roger.

With a flourish, he served.

"Would you like an ice-cold Genesee Malt Ale?" Roger handed Tom and me chilled bottles. I took a slow, deliberate sip and closed my eyes.

When I returned to reality, Roger had already carried bowls and platters to the table, among them real-meat creamed chicken with mushrooms and green peppers on rice, whole wheat bread, corn on the cob, tropical cake, and tea. We relished every morsel and ate every crumb.

Vignette 128: Who Is Roger?
Tuesday, June 26, Day 20

Oh, taste and see that the Lord is good! Blessed is the man who takes refuge in him

(Psalm 34:8 ESV)!

As we ate and ate, we discovered more about our host. He insisted on a 5:30 supper so he would be free to watch his favorite show at 7 p.m.—MacNeil and Lehner's news. Roger was a retired teacher who had taught social studies in Queens, New York.

He explained that he began opening his cottage to hikers in 1973 and has become a legend on the AT, even featured in Sports Illustrated for his efforts. We promised ourselves we'd find that article someday. We did. (July 13, 1981, page 78, Title: They're beating a path to the door of Roger Brickner's hiker's haven—Appalachian Trail)

While Tom and I did dishes and Carla read magazines, Roger nestled down in his dilapidated recliner to watch his show.

"Hey, Mom, we're going to the lake to fish," Caleb called over his shoulder as all three boys swung out the door.

I glanced at Tom. "They already walked seven miles on the AT, another mile up from the lake, and now they're hiking another almost two miles roundtrip to fish?"

"Is it the lure of fishing or the joy of walking?" Tom shook his head.

Later, all of us together, Roger showed us his computer, set up to record weather observations. Weather is another of Roger's interesting hobbies. He shared a video of his February summit of Mt. Washington where, according to the T-shirt he sported, they have the worst weather in the world.

Roger took several short calls throughout the evening concerning the sale of his cottage. We were sad to hear that Roger's cottage would soon be history on the AT but happy to hear of his plans. He was buying an 1810 home on the green in Haverhill, New Hampshire, for $325,000. Since that little borough was only nine-and-a-half miles from the AT, Roger planned to make a daily run to the trail, picking up hikers at a designated time and place. He'd treat them to the same hospitality we enjoyed here.

By the time Roger brought out the rocky road ice cream, Carla had dozed off on the floor, not to be roused, even for ice cream. The rest of us packed it in before crawling into our nests for the night—the boys on mattresses on the floor in the front room and Tom, Carla, and I on single beds on the screened back porch. Peace.

~Distance hiked on day 20: 6.5 miles

~Wawayanda State Park, New Jersey, to Roger's Cottage, Greenwood Lake, New York

Vignette 129: Burden Free
Wednesday, June 27, Day 21

Blessed be the Lord. Day by day he bears our burdens
(Psalm 68:19 EHV).

I contributed my blueberries and pancake mix to breakfast. And since I remembered to soak the berries over night and since I didn't dump them into the mix, water and all, and since we had an actual griddle on a stove, we had fluffy blueberry pancakes instead of mush. Roger complemented our fare with bananas, orange juice, sausages, coffee, and warm conversation. Decades later, remembering breakfast at Roger's cottage still fills me with warm feelings of all-is-well.

With dishes done and a donation left on Roger's windowsill, we reluctantly packed up. But Roger offered one more kindness. "Hey, would you like me to carry your backpacks to Orange Turnpike? I'm going there anyway to fill my twenty-six water jugs at the spring. Some thru hikers work out walking parts of the trail with no backpacks, slack packing. Helpers move their packs ahead."

"I'm not opposed to slack packing," Tom said. "We are definitely not purists in hiking the AT."

"But that's fourteen miles down the trail. What if we can't make it by 4 p.m.?" I doubted we could cover the rugged terrain that quickly and did not want to inconvenience Roger for one second.

"We can do it no problem, Mom," Caleb piped up.

When I studied those three healthy, trail-toughened young men, I knew he was right. Even if Tom, Carla, and I weren't at the spring to claim our packs, the boys would be.

We repacked supplies, taking only lunch, raincoats, wallets, and first aid kits and divided everything between two light-weight packs, the two that wouldn't fit in Roger's small car. At 8:40, we started the steep climb from Roger's cottage. As usual, we broke into a sweat, but only a slight sweat. Having no packs or feather packs on our backs was dreamy.

At the top, in high spirits, we allowed Carla to hike ahead with the boys. Her face lit with joy at being considered worthy to hike with the big kids. Just moments after we'd agreed on a meeting place, they were gone.

Freed of my heavy pack and with a fresh supply of caffeine zinging through my veins, I strolled contentedly along. I felt like we'd just had biscuits and coffee at Hardees, a stop on a regular walk for us, and that we were now strolling home. Only our views were considerably more magnificent. Greenwood Lake sparkled far below and blue-colored mountains paraded in the distance.

Sometimes as we walked, we held hands. Sometimes we discussed our future plans. "I really like no weight on my back." I grinned at Tom.

He grinned back. "Hiking without the burden of our packs is a breeze."

Vignette 130: Giving Thanks
Wednesday, June 27, Day 21

May all who seek you rejoice and be glad in you! May those who love your salvation say evermore, 'God is great'

(Psalm 70:4 ESV)!

"Let's have a party for all the people at home who supported us on this hike," Tom suggested.

"Good idea."

And so we began to create our guest list in Tom's little notebook.

"Our neighbors, Karen and Rob, have been checking our house all this time," Tom said.

"And you even asked Karen to throw down some grass seed by the swing set and take your film in for developing," I added. "Yes, they definitely need to be invited."

"Remember our neighbors, Ray and Arlene, came to our rescue the morning we were departing? We were trying to fit six packs and six people into our station wagon and discovered we'd have to leave two packs or two kids home," Tom said.

"Yep. We were scratching our heads in the driveway when Ray came out his front door and offered his car-top carrier to us for the summer."

We laughed at how we hadn't even considered that we wouldn't fit into our car until that moment and how God took care of us despite our ignorance.

"Greg and Lisa deserve an invitation." I smiled. "She's been shipping our food boxes, and both of them have written notes of encouragement."

"We can't forget Dave and Rosie," Tom said. "I'm using his pack."

"And I'm sleeping on his pad. And what about Paul and Char? They loaned me the food dehydrator, recipe book, and lots of encouragement."

"Remember when Lois shared her backpacking expertise with me?" I asked.

"Yes, Ray and Lois make the list," Tom agreed.

"Louise and Jim have been dog sitting Cody this whole time. Let's invite their family."

"Let's also invite Jerry and Patsy for all the uplifting letters and the care packages they sent."

We continued to add people to our list and then chuckled over some of the menu items we'd serve. "Be sure to serve everyone nut mix and dried apricots," Tom said.

"Okay. Let's also serve a basketful of granola bars. And we must include Hearty Hungarian lentils in pita bread and a dish of Spanish rice with almonds."

Tom grimaced, glad he could bypass those nutritious foods for a good-old, cholesterol-packed hamburger and a sugary can of soda.

"Thank God that we were blessed by all these kind people."

Vignette 131: Easy-Peasy
Wednesday, June 27, Day 21

For with you I can charge against a battalion, and with my God I can jump over a wall

(Psalm 18:29 EHV).

We zoomed down the trail and had covered almost four miles by 10:30 a.m. when we came to the end of Section 13 of New York's AT and began Section 12. As planned, the kids were there. "We've been waiting forever," Ben whined.

"Okay, go!" Tom swung his hand down the trail. "We'll see you at Fitzgerald Falls." And they were off.

But Tom and I paused at the summit of Eastern Pinnacles, absorbing the majestic views north, east, and then south over Greenwood Lake Valley. I studied that lovely valley and took a deep breath. "I hate to leave it, but Graymoor Monastery is beckoning. Can you believe we're within a couple days of finishing our hike?"

"I'm not sure how I'll feel when we've reached Graymoor," Tom said, "but I think I'll be rejoicing."

Shortly, we reached the steep climb over Cat Rocks. "I'm thankful we don't have heavy packs on." I panted and clung to rock and roots. "And I'm glad the kids have already climbed this. Otherwise, I'd be worrying and fretting, yelling cautions to them as they fearlessly scampered up the cliffs."

"We're all glad," Tom said, between gasps.

Two miles beyond Cat Rocks we met up with the kids at the base of Fitzgerald Falls. This twenty-five-foot waterfall cascades off rocks and offers interspersed, multi-level wading pools. There, calmed by the tumbling water and comforted by the natural foot-soak tubs, we lunched on cheese spread, crackers, fruit ripples, and, of course, granola bars.

After our meager picnic but lush setting, we set out again, this time keeping Carla with us. Again, the AT climbed steeply, straight up the cliff to the top of the waterfall. As I planted each foot and clutched roots, my unnecessary litany began. "Be careful of this slippery rock, Carla. Keep a firm hold."

Vignette 132: Not So Easy
Wednesday, June 27, Day 21

For he will deliver you from the snare of the fowler and from the deadly pestilence. He will cover you with his pinions, and under his wings you will find refuge; his faithfulness is a shield and buckler

(Psalm 91:3, 4 ESV).

The humid eighty-six degrees and the many miles to hike made for a long afternoon. At 9.7 miles, we reached Mombasha High Point.

"The Philosopher's Guide says we should be able to glimpse the skyline of New York City to the south," Tom said. The three of us strained to see something, anything, through the haze. Nothing.

"We'll have another chance to see NYC from Black Mountain or West Mountain." Tom referred to the book again. "But see over there to the east is Harriman State Park. We're going there."

I was reading the guidebook. "Before Harriman, we get to cross several streams and many jumbled rocks, and two summits of Buchanan Mountain."

"It's always a pleasure when a mountain has two summits." Tom shook his head in disbelief.

"Don't forget the jumbled rocks." Carla giggled and skipped off.

Jumbled rocks were a major part of our afternoon hike. Rocks ranging in size from basketballs to pianos were scattered in piles. The AT careened over and around them. Carla and I nicknamed this part of the hike "jumbo jumbled rocks jumble" and then tried to say it ten times fast without mixing it up. The game took our minds off the escarpments, steep ascents and descents, and finally the second summit of Buchanan Mountain.

From there we could see Little Dam Lake and, in our responsible adult roles, Tom and I refrained from any jokes about its name and its relationship to the AT. Shortly we crossed an inlet of Little Dam Lake on a wooden truss bridge, lovingly built by a hiking club in the area, and we regretted our nasty thoughts about the AT.

But not for long.

Horseflies and jumbled rocks took turns being pesky the rest of the way to Orange Turnpike. Nasty thoughts about the perils and irritations of the AT returned.

Vignette 133: Not Like Home
Wednesday, June 27, Day 21

By day the Lord commands his steadfast love, and at night his song
is with me

(Psalm 42:8 ESV).

Carla, Tom, and I arrived at the spring at 4:15 p.m., just a half hour after the boys and only five minutes after Roger. We gulped down fresh, ice-cold spring water coming from a pipe in the hillside and formed a brigade line, filling Roger's jugs.

"Well, that's the fastest I've ever filled twenty-six jugs. Thanks!" Roger shook all our hands and jumped into his little car. We stood forlornly, watching his car disappear around the bend.

"Wouldn't this be a convenient spot to camp, next to fresh water?" I asked.

"Let's see if we can find two flat patches for our tents." Tom nodded and scanned the area while the rest of us kicked about, up the hill, around the swell.

"Nothing." I sat on a rock. "But we could eat supper here."

"Let's do it, and then we can hike further," Joel said.

We cooked stroganoff and topped it with almonds and downed quarts of ice-cold water. Then we shrugged into backpacks that had gained weight while taking their day off. After lumbering down the hill onto the edge of the turnpike, we plodded back the half mile to the AT.

"It's only six o'clock. Let's hike another three miles or so yet today," Joel suggested, as we entered the trail.

"Yeah, then we'll be closer to the monastery," Caleb agreed.

"And closer to being done," Ben mumbled.

With determination in their steps, the boys began the immediate cliff-like ascent of Arden Mountain with Carla and me not far behind.

"You boys wait at the top," I yelled before they were out of sight, over the next mountain.

But Tom's stamina faltered. By the time he reached the top of the half mile climb, where the rest of us perched on rocks, Tom made an executive decision. "We will camp here tonight." Ignoring groans and pleas, he set up camp. Reluctantly, the rest of us did too.

Actually, it was a fabulous place to camp. There were awesome views of the sun setting over Mombasha High Point and Bellvale Mountain with the tips of the Catskill Mountains visible to the north. We had giant slabs of rocks to sit on, soft moss beds nearby to cushion our tents, and a big fire pit. The boys made a fire. We took off our shoes and socks, lining them up near the fire to dry.

Caleb and Joel took turns popping corn. We leaned on our packs with contentment, passing around the popper. Before the popper got to me, every time, only the leftover kernels remained. I nibbled on the seeds.

"I'm making one more batch of popcorn for Mom," Joel announced. When the vultures, otherwise known as my children, approached, Joel shooed them off. "This is for Mom." Love was delivered via popcorn.

Vignette 134: Longing for Home
Wednesday, June 27, Day 21

Even the sparrow has found a home, and the swallow has found a nest for herself, where she may place her young near your altars

(Psalm 84:3 EHV).

At dusk, Ben and I hiked back to the spring to refill our canteens. It was a little less than two miles round trip. We knew the trail ahead was dry, and we wanted to be sure we had enough water.

"I'll be really glad to get back home," Ben confided. "I know my friends are having fun going to baseball games and hanging out together, and I'm stuck here."

"I know, Ben." My heart hurt for him. "But when your friends don't have a clue what they did during the summer of 1990, you'll remember with great detail, and maybe even fondness, that you were hiking the AT with your family."

He shrugged.

When we returned to camp, the rest were finishing up a game of Bunco. Joel and Caleb lay down on the flat rocks, watching the stars twinkle, and discussed their philosophies of life. I listened in the background.

"I plan to live a reckless and full life," Caleb stated.

"I think you should be careful and live a long life," Joel said.

They chuckled at their differences.

Just then, Mountain Mist hiked into our camp. "Hi, folks." His eyes scrunched into a grin as he pulled out his one-man tent.

"Hi, Mountain Mist," we said in unison, surprised to see him again.

"How did you get behind us?" Tom asked.

"I took a couple of days off and went home."

Home. What a sweet-sounding place. Homesickness engulfed all of us. As the boys climbed into their tent, Caleb and Joel were still conversational and discussed what they would be doing at home.

"I'd be playing basketball in the backyard or football in the field," Caleb said longingly.

"I'd be in the basement work area, fixing a radio or rewiring my walkie talkie," Joel mused. "I wish I were there right now, figuring out a way to destroy the AT."

Ben joined their raucous laughter, and moments later all was quiet, except for the sounds of a gentle rain moving through.

~Distance hiked on day 21: 15.4 miles (17.4 miles for Ben and me)

~Roger's Cottage, Greenwood Lake, New York, to Orange Turnpike

Vignette 135: Danger Ahead
Thursday, June 28, Day 22

When evildoers advance against me to eat my flesh, when my foes and my enemies come against me, it is they who will stumble and fall *(Psalm 27:2 EHV)*.

"Let's get an early start." Instead of packing up, however, Tom launched into his medical ritual—applying moleskin to the blisters on his feet, wrapping his knees for stability, and greasing the areas on his sides that were rubbed sore from his pack.

"Good idea." I knew I had a few minutes of quiet time, so I moseyed to the ledge that overlooked Orange Valley and wrote in my journal.

A thin veil of fog rested on the treetops below me, but the sky above was clear. Out of sight, the traffic on Orange Turnpike swished along. I contemplated our hike. We'd been on the AT three weeks. We'd shortened our goal of 500 miles to 250 miles. And now only twenty-eight miles remained to Graymoor Monastery. I wondered if we could push ourselves and do it in two fourteen-mile days. I wasn't in favor of pushing ourselves beyond our limits just to finish one day earlier. It seemed wiser to enjoy the walk and spread it out over three days in a more manageable nine miles per day. If the climbs were too steep or the trail too rocky, the decision would be made for us.

Yet I longed—we all longed—to be done with the hard work and discomforts of the AT. Leaning over and scratching three new bug welts, I was reminded of one of those discomforts.

"Let's go!" Tom's yell shook me from my scratching reverie. Soon the kids were tumbling out of the tents. Thanks to our well-tuned team effort and a quick meal of breakfast bars, we set out by 7:10 a.m.

"I think we can make it to West Mountain shelter today," Tom announced, revitalized from a good night's sleep. His enthusiasm fired up the kids.

"Yeah, let's do it!" Ben exclaimed.

"Yeah," the rest cheered, and the boys roared out ahead.

Only I remained quiet, groaning to myself, knowing we had sixteen grueling miles to reach that goal, knowing that I'd once more be stuck in the middle of a tug-o-war. The kids would surge ahead. Tom would lag behind. I'd be somewhere in the middle, worrying about both ends of our lineup.

West Shelter also held a fearful scene for me . Last year a hiker had been shot there. The attacker had never been caught. Hiking near there seemed frightful, staying there seemed insane.

Vignette 136: Agony and Ease
Thursday, June 28, Day 22

O God, you are my God; earnestly I seek you; my soul thirsts for you; my flesh faints for you, as in a dry and dreary land where there is no water

(Psalm 63:1 ESV).

I put fear aside and turned to the first obstacle of the day, the descent of Agony Grind. A steep descent is no problem if your joints are lubricated and your muscles can handle isometrics. Descending with gravity on your side doesn't take the stamina of ascending either, but ask older hikers which they prefer and they'll want an ascent any day.

We hadn't considered bringing hiking poles on this hike. We were young, surefooted, and strong. Hiking poles were for others, not us. In looking back, after years of hiking with poles, I see the folly in our thinking. Poles help you go up and down, plus add stability on wobbly rocks. But we didn't know that yet.

With quiet dread, already limping, Tom approached Agony Grind. As the path dipped over the crest, we found a walking stick, complete with a rubber bottom, leaning against a tree. Attached to it was a note.

Dad, I found this along the trail. Maybe it will help! Joel.

"How thoughtful," I said. "It's like a gift dropped from heaven."

"I don't need help." Tom stared at the stick.

"It might be worth a try." I handed it to him.

Tom planted it on the ground. "Maybe leaning on this walking stick will help my knees." He took a careful step down. "This might just work." He turned back with a grin.

And it did. By 8:30 a.m. we had conquered Agony Grind and were ready to enter Harriman State Park. The first section of the AT was opened here in 1923. To celebrate this momentous occasion, Tom, Carla, and I took a twenty-minute break, letting the traffic whiz by on

Highway 17 while we snacked on raisins. I wondered where the boys were and pondered Tom's condition. He seemed so breathless, his energy already sapped.

At first, the hike through Harriman was a snap as we strolled on paved Arden Valley Road and then crossed the New York Thruway on a railroad bridge. Next we wound through scattered grassy campsites and then began the exhausting ups and downs hikers expect in Harriman.

Tom continued using his new hiking stick. His knee pain was eased, but his gasping for breath continued.

Later we learned that Tom has a heart condition known as atrial fibrillation.

Vignette 137: Crazy Falcon and Tight Squeezes
Thursday, June 28, Day 22

Light dawns in the darkness for the upright

(Psalm 112:4 ESV).

Soon we came to the landmark known as Balanced Rock. A fern-topped boulder towering above our heads was held in an upright position by a small stone propped under one side. What amazing act of God. I was still staring at the boulder when Carla, hiking several yards ahead of us, screamed and threw herself on the ground. A falcon swooped over her head.

Carla crawled back to us, sobbing. "I could hear her wings swish over my head."

I hugged her and peered into the sky, expecting another attack. "She must have a nest nearby and is protecting her babies. Let's get low and get out of here."

The boys were waiting for us at our designated meeting point, the Lemon Squeezer, a narrow, steep passage between large boulders.

"We'll have lunch here before we attempt squeezing through," I said.

"Under normal circumstances, I'd think that was a silly idea," Tom said. "Why would we want to eat lunch before a tight squeeze? But then a few sunflower seeds, fiber bars, and dried apricots won't make us too fat, will they?"

Soon the boys were plunging through the Lemon Squeezer without hesitation. I watched them disappear into the darkness. Next little Carla skimmed through the thirty feet of threatening boulders, and I saw her emerge into the bright sunlight on the other side. I stood analyzing the skinny ravine, wondering if I'd be narrower facing forward or sideways with my backpack attachment. Dragging the pack behind wouldn't work. Taking a detour on the blue bypass trail around the Lemon Squeezer was out of the question. I didn't want to be a coward. Holding my breath, I plunged in.

With rock walls pressing in on both sides, I scraped along, pushed lower and lower as the rocks grew darker and tighter, until I was in a squat and then a crawl. Only the black ground filled my vision. Slowly the ground grew lighter, and I emerged into the bright sunshine. Still in a squat-crawl position, I flipped helplessly onto my back like a turtle with a pack on. Tom, who had detoured around, was there to set me upright, and Carla beamed proudly down at me. The boys, of course, were out of sight.

Tom's journal entry reads, "I was initially going to attempt the Squeezer also. Carla made it through and so did Janie, but…" There Tom's journal ends forever. It was not the end of Tom. He took the blue blaze trail.

Vignette 138: Reprieve
Thursday, June 28, Day 22

He makes me lie down in green pastures. He leads me beside still waters. He restores my soul

(Psalm 23:2, 3 ESV).

The three of us crept along, tired and sticky, through the bleak swampland of Island Pond Mountain. At Tiorati Circle, the boys waited, pouting, on the side of the road.

"We've been here an hour." Ben shook his head.

"We're starving! Could you cook a hot meal?" Joel begged.

"Now?" I was depleted. I stared at the nearby sparkling lake with a beach, a shower house, and a faucet. Fresh water was there for the taking, but, most importantly, I saw a patch of soft green grass where I could collapse for a half hour.

"Sorry, boys, I'm not cooking right now. But one of you is welcomed to."

"I will." Joel jumped up and dug through our food bag. "I think we'll have tuna cheese broccoli rice and Malt-o-meal for dessert."

"And I'll help," Caleb announced as he set up our portable kitchen.

I nodded with satisfaction and collapsed on the soft grass while the rest of the family made lunch.

Thirty minutes later, Joel called, "Come and get it, Mom!"

He had been right. We all needed a hot meal.

"Delicious!" I grinned at the kitchen crew.

"I'm sorry that the Malt-o-meal is a little burnt," Caleb said.

"Let's just disguise it with extra brown sugar." Tom scooped several spoonfuls into the kettle, and we slurped it down for dessert. Since I was still on strike, the boys did the dishes.

"What do you say we play in the lake for awhile?" Tom asked.

"Really?" the kids chimed together. This wasn't their goal-driven dad's normal schedule. We all set off for an hour of fun.

The beachgoers, in their fancy swimsuits, didn't mind our plunging into the lake in our t-shirts and shorts nor our oohing and aahing as we sank into blissful oblivion. To be immersed in cool water, to dunk one's head and let the water fill all the sweaty cracks and crevices, to float lazily along on a carpet of cool velvet are all very close to heaven. After we'd adequately cooled off and then showered, we were well-fed, refreshed, and ready to go.

Vignette 139: Friend Support
Thursday, June 28, Day 22

If I say, 'Surely the darkness shall cover me, and the light about me be night,' even the darkness is not dark to you

(Psalm 139:11, 12 ESV).

While Tom taped himself back together, the rest of us lounged at picnic tables. Two thru hikers came along—Olive Oyle and Brutus. I was impressed, as I was with every thru hiker, knowing they'd come all the way from Georgia on the AT. And as all hikers do, we discussed what we had in common—the perils, hardships, surprises, and joys of the trail.

"I was attacked by a falcon back there." Carla pointed down the trail.

"That falcon dive bombed Olive Oyle too." Brutus chuckled. "I was smart enough to let her go first."

"Very sneaky." Olive Oyle gave him a shove. "Nobody told me about that falcon, that it attacks most hikers, especially the first in a line of hikers. You're very brave, little girl." Olive Oyle grinned at Carla.

Carla stood straighter.

Before Brutus and Olive Oyle left, they gave us a tip on a shortcut back to the AT. Soon we stepped off with hopes of making it to West Mountain Shelter that night. Only I struggled with the dread of sleeping in the crime scene.

The hike to Brien Shelter skimmed by. There we met Curt, another thru hiker, and caught up with Brutus and Olive Oyle. The couple continued ahead, but we chatted with Curt a few minutes. He was having a snack of a box of cake mix and water. Not a healthy snack but filling and calorie loaded.

We didn't loiter long, even though I wanted to discuss the benefits of healthy snacks. It was 7 p.m., and we hoped to make three more miles. We hiked frantically, but an hour later at Palisades Highway, I had the sickening realization that we still had one-and-a-half miles to go on the trail—most of it uphill to the top of West Mountain—and then another six-tenths of a mile to the shelter.

Dusk was closing in.

As we began our ascent, Curt caught up with Carla and me. The boys were already far ahead, and Tom was struggling far behind. As I worried aloud, Curt assured me that it would not be as dark on the top. At confusing intersections or treacherous steps on the trail, Carla and I waited for Tom. So did Curt. But every wait brought total darkness closer.

Vignette 140: Worrisome West Shelter
Thursday, June 28, Day 22

Guard me, O Lord, from the hands of the wicked; preserve me from violent men

(Psalm 140:4 ESV).

Finally, with only a hint of light left, Carla, Curt, and I reached the crest of West Mountain. With relief, I spied the boys. They had already scouted out the area.

"It's too far to go to West Mountain Shelter," Ben announced. What brilliant children. I nodded.

"We can set our tents here by the fire pit." Caleb stood by it on a level patch.

"Looks like you have everything under control." Curt tipped his head to the boys and hiked on.

Too tired to talk but not to worry, I dropped my pack and walked back down the trail, straining to hear any movement below. Not too far down, I heard Tom heaving himself from a rock to a tree to a rock to a tree and then a rhythmic plod. I turned back, not wanting Tom to see my concern.

In total darkness, we set up our tents. We didn't even need flashlights to do the now-common routine. Our canteens were empty again, but we feasted on Spam and crackers, plus two cans of soup that were left near the fire pit by someone who couldn't tolerate the thought of carrying an extra pound one more step. God bless them.

As the kids settled into their tents, Tom said to me, "Let's try to see New York. It should be visible from the edge."

It was an exciting thought, so we tripped and scrambled to the edge of the mountain, facing NYC, hoping to catch a glimpse of the glowing skyline of that thriving metropolis, so near to us yet a world away. It was lost in the haze.

Disappointed, we stumbled back toward our tent. Six-inch rocks, strategically hidden in twelve-inch grass, continued to batter our feet.

"I feel like my feet are bruised all the way through." I winced with each step until we scooted into our tent and wiggled into our bags.

"What a relief to be on top of West Mountain and not at the shelter," I said, as if that half-mile cushion could save us from a murderer.

Tom snored.

~Distance hiked on day 22: 16 miles

~Orange Turnpike, New York, to West Mountain Summit

Vignette 141: Water Boys
Friday, June 29, Day 23

I am worn out from my crying. My throat is sore. My eyes are blurry,
as I wait for my God
(Psalm 69:3 EHV).

Water was our first concern of the day. Since we had gone to bed with none, we had none to drink or even to make oatmeal. According to the guidebook, a spring was six-tenths of a mile away, near West Mountain shelter.

Yes, the dreaded West Mountain shelter loomed up to haunt me again.

"Caleb and I will go get water," Ben announced. As I squelched my fears, they took off down the blue-blazed trail toward West Mountain shelter with canteens and kettle in hand.

The rest of us packed up camp and put out breakfast. Forty-five minutes passed. "Don't you think Ben and Caleb should be back by now?" I peered down the trail.

"Nah, it will take them at least an hour." Tom shrugged and worked on his morning medical rituals. Joel started a fire in preparation for heating water for our oatmeal and hot chocolate. To keep busy, Carla and I picked up bits of trash and tossed them into the fire.

More than an hour passed. Still Ben and Caleb didn't return. By now, last year's shooting at West Mountain shelter dominated my thoughts, and I paced. "How could I have sent two of my boys in that direction? What was I thinking?" I paused in front of Tom.

"They're fine." Tom settled on a rock.

"You don't know that. I hate when you say that."

I glanced again at the blue-blazed path. It lured me down, and I began to hike. The trail was desolate and parched. Joel and Carla crept behind me. Every few feet, we paused, shouted, and listened.

"Ben! Caleb!"

"Please, God, let them be okay," I whispered.

With every silence that followed our calls, my panic mushroomed.

After several minutes of walking and calling, a disgruntled reply came. "Yeah, yeah. What's all the yelling about?" Ben stomped up the trail. "There's no water here. We must have walked four miles, and all we found was this."

Caleb thrust out the kettle, showing only a couple inches of muddy water in the bottom. I swooped upon them with delight. They were alive. No bullet holes. They stared suspiciously at me, convinced the trail had finally damaged my brain.

Back at camp, Tom tried to filter the thick water with our retired pump. When he had enough for a couple of swallows each to wash down a dry granola bar, we dumped the rest and started on our long day of hiking.

Vignette 142: Water Worries
Friday, June 29, Day 23

Call upon me in the day of trouble; I will deliver you, and you shall glorify me

(Psalm 50:15 ESV).

It was already 9 a.m., not a good starting time for reaching Graymoor Monastery, still twelve miles away.

"We have to get to the monastery by 5:30 p.m.," Ben announced. "That's suppertime, and no latecomers will be served."

"But we need a meeting point before that." I gave Ben a stern look. "Let's meet at the zoo, about five miles ahead. The guidebook says we'll find water there."

Ben's shoulders slumped. "Okay, but don't take all morning."

"We don't want to stop sooner, either. We all need water." Tom managed a grin.

The one-and-a-half mile walk down West Mountain captivated us with beautiful views to the east and southeast over Bear Mountain. The broad waters of the Hudson River unfolded below. Part of the hike near the bottom of West Mountain took us on the route of the 1777 Trail. An hour passed.

Tom, Carla, and I paused to snack on raisins and almonds. "We can see water, but I need to have some." Tom shook his head. "I'm not sure I can make it much farther."

Carla and I exchanged worried glances. We didn't need as much water as Tom did. In the same amount of time that Tom would guzzle a two-quart canteen, Carla and I would share one-quart.

We began the next ascent—Bear Mountain.

"See you at the top," Carla called out to Tom. We pulled ahead, but about a mile up the trail, at a tricky intersection with paved Perkins Drive, Carla and I waited fifteen minutes for Tom to catch us. We worried that in his water-deprived stupor, he might miss the trail.

As our feet pounded up the paved road, Carla and I prodded Tom with encouragements.

"It's only a little way to the top." I patted Tom's arm.

"The hike down will be simple." Carla smiled at him.

Tom shuffled and gasped.

Carla and I walked and waited, walked, waited.

"Take one step at a time." I fought off impatience. The monastery was still nine plus miles of rigorous ups and downs ahead, and I was desperate to be there in time for supper. I had to be there in time for supper. Really had to.

Vignette 143: Life-Saving Water
Friday, June 29, Day 23

Then he led out his people like sheep and guided them in the wilderness like a flock. He led them in safety

(*Psalm 78:52, 53 ESV*).

I felt like I had a bungee cord tied around my waist, attached to a heavy weight pulling me backwards. I needed to distance myself from that heavy weight, from the obstacle that was keeping me from getting to the monastery for supper, from Tom. I also wanted to save our marriage by avoiding any thoughtless words. When the trail left the road and reentered the woods with Bear Mountain summit only a half mile ahead, I said to Tom, "We're going to hike ahead."

"See you at the top," Carla called back.

The top was better than we imagined. Heaven awaited us at the top. We emerged from the trees and there, glowing brilliantly in the sun, the highlight of the scenery, was a drinking fountain. Next to it perched one of our canteens, filled with water, signaling that the boys had safely passed through.

"Water!" I squealed with delight at the sight of a civilized water source. Carla and I dashed for the drinking fountain, but a little girl was there first, slurping water, letting it run off her chin and down the drain. I wanted to scold her for wanton wastefulness. I wanted to push the little girl aside, but instead I got control of myself and waited my turn.

The cool water refreshed my parched throat. My dehydrated body absorbed the liquid. My brain could focus again.

With filled canteens, Carla and sat on the crest, waiting. "Tom, we have water." I held the sparkly plastic bottle over my head, sloshing it, hoping Tom could see and hear us.

"Really?" A faint but hopeful reply sounded.

Soon Tom was gulping water, dumping water over his head, and gulping some more. We laughed in relief. We thanked God for the blessing of water.

Finding that pump was a definite turning point in our day, turning us from despair to gladness. And later, when we read in the Philosopher's Guide, "Water bubbler and restrooms at the summit are open seasonally," we laughed at our stupidity. Why hadn't we checked that other reliable guidebook?

Vignette 144: Inspirational Bear Mountain Bridge
Friday, June 29, Day 23

You have given him his heart's desire and have not withheld the request of his lips

(Psalm 21:2 ESV).

We caught glimpses of the Hudson River far below as we began our descent.

"We must be getting close to Bear Mountain Bridge," I said. "We need to cross it on our way to the monastery." We peered through the trees and plodded along, concentrating on the trail, occasionally glancing at the river.

Carla skidded to stop. "Look! I see it." We stared down in the direction of her pointed finger. There was the girded span, the highway bridge, over which the AT passes.

"I can't believe it," Tom whispered. "That's the picture we've been seeking on this never-ending hike."

I put my arm around Tom's waist, feeling the magnitude of this moment. "Ever since we first saw a photo of Bear Mountain Bridge in *Readers Digest*, we've dreamed of some day hiking over it. Sometimes it seemed impossible. Now that dream is about to come true."

"Let's go!" Carla shook us back to her reality.

Tom snapped many photos as we traipsed down the trail. More to add to an already great pictorial record of our hike.

At the bottom, at the entrance to the zoo, the boys were fishing in a pond and waiting. Of course, we were one-and-a-half hours behind them.

"Did you guys even see Bear Mountain Bridge?" Tom asked, exasperated with their break-neck speed.

"Yeah, yeah." Ben dismissed his enthusiasm.

"What's so great about another bridge?" Joel challenged. "I think log bridges are cooler."

"Let me tell you about Bear Mountain Bridge." I stood with hands on hips, my teacher-mom role showing. "When Bear Mountain Bridge opened in 1924, it was the longest suspension bridge in the world, the first bridge to span the width of the Hudson this far south, and the first to use a concrete deck."

I had their attention. "Did you know that construction techniques on it were applied to build bigger suspension bridges, like the George Washington in 1931 and the Golden Gate in 1937? It was and is an engineering marvel."

"Okay, Mom, is that all?" Caleb shifted from one foot to the other. "When can we eat?"

I shrugged and pointed to a table, then to each of my three sons. "Let's eat here. But after lunch, before you take off in your hiking frenzy for the monastery, pay attention to this: you'll be hiking over a bridge that in 1982 was added to the National Register of Historic Places."

Tom stepped in. "That picture of this bridge inspired us to do this hike."

"Wow. I am inspired," Ben muttered. Tom and I shook our heads while the kids slapped each other on the backs and laughed.

Vignette 145: Deadlines
Friday, June 29, Day 23

I shall walk in a wide place, for I have sought your precepts

(Psalm 119:45 ESV).

It was 12:30 p.m. when we had our oatmeal and hot chocolate. While we feasted on our unusual lunch, Olive Oyle and Brutus hiked up.

"I always do love breakfast for lunch." Brutus laughed at our meal, but they didn't stop long to chat since they had to pick up a food shipment at the post office in the nearby town of Fort Montgomery.

"We need to hurry. Latecomers will not be served supper at the monastery." Olive Oyle shook a finger at Brutus.

Brutus took off, heading up hill, over the embankment. Olive Oyle paused, glancing down the sidewalk at a white blaze.

"Where are you going?" Caleb asked Brutus.

"The trail is on the sidewalk." Joel pointed down the for-once-easy trail.

Brutus backtracked, chuckling. "After hiking the AT from Georgia and following its difficult paths, I never expected the trail planners to pick the easy way."

At 1 p.m., we were ready to burn trail too. We had a little over seven miles to go to the monastery and four-and-a-half hours to do it. On a good day, we could handle two miles per hour. We had a chance.

But the boys were taking no chances. "Bye, see you at the monastery," Ben said and led his brothers in a near-trot down the sidewalk through Nature Zoo.

The Philospher's Guide had some interesting thoughts here, inviting us to truthfully view wildcats and bears during our hike. It also cautioned us that we may be the spectacle among the zoo viewers and even feel a little intimidated by the big city tourists.

But we did not feel low or intimidated by the crowds. We'd already been low and intimidated for weeks. We did chuckle, though, to see the familiar white blazes on the sides of buildings, pointing AT hikers on a meander through the zoo.

We emerged, unharmed from our fling with wild animals.

Vignette 146: Wrong Turns
Friday, June 29, Day 23

By day I called, and you answered me. You have made my soul strong

(Psalm 138:3 EHV).

"Here's the challenge, Tom. If we don't make it to the monastery in time for supper, I can always cook up the rest of the lentils and brown rice for us."

He wrinkled his nose. "We'll make it." And he sprinted off in Ben-fashion.

Crossing the Hudson on Bear Mountain Bridge was exhilarating. Its two lanes still carried the traffic of US 6, US 202, the Appalachian Trail, and State Bike Route 9.

"Look how long it is." Carla squinted up at its 2,255 feet span.

One-hundred-fifty-five feet below us, a barge passed, then a pleasure boat, while the highway traffic zoomed along.

"We're here, doing this." Tom grinned.

Brutus and Olive Oyle scurried back toward us. "We thought the town and post office were on the other side of the river." Olive Oyle's eyes brimmed with tears. She glared at Brutus.

I could imagine she was thinking that it was his fault that they would be eating noodles or rice with lentils for supper when they could have been feasting at the monastery. Brutus may have been thinking the same about her. Neither one had anything else to say to us or to each other as they stomped back. Oh, the relational strains of hiking the AT together.

One must be persistent, forgiving, patient, consistent, brave, stouthearted, strong, and much more to reach this point—especially if you hiked from Georgia, especially if you hiked with a family of six.

Vignette 147: Friendly Support
Friday, June 29, Day 23

May he grant you your heart's desire and fulfill all your plans

(Psalm 20:4 ESV)!

Ahead of us loomed our last obstacle. It was a mountain called Anthony's Nose. I wondered if it was a hook nose or a pug nose. I expected the hook and got it.

A half mile after the bridge, we began a gradual ascent, followed quickly by a sharper one. One hiker had said, "Pick your way up Anthony's Nose." And we found out why. There was no walking, just a step-by-step, pick-your-way-up as the trail rose to eye level only a few feet ahead of us. It was a grand finale to two dozen days of toil and sweat on the trail. We were grateful that this steep challenge only lasted a half mile.

Near the top we came once more to the familiar sign, *Welcome to the Appalachian Trail. Open to all who walk.*

"I think we should say open to all who sweat," I said.

"Remember Boz?" Tom asked. "He told us that showers and laundromats last about a half hour on the trail. Haven't we grown accustomed to feeling and smelling sweaty?"

I sniffed my salt-ringed arm pit. "Speak for yourself, Sweat Hog."

"I heard that." Curt smiled down at us as we crested Anthony's Nose.

"What are you doing here?" Tom looked up. "You should be miles ahead of us."

"Stopped for my food shipment and lunch in a restaurant," Curt said. "Mind if I keep you company?"

"I would appreciate it." Tom nodded.

It helped Tom forget the pain of these last hours of forced hiking. Curt, twenty-nine years old, was a pleasant fellow. He quit his job to spend six months hiking the AT—a fulfillment of a dream—and was quietly satisfied with his life at the moment.

A couple months later, we received a photo of Curt on top of Mt. Katahdin—dream accomplished!

Vignette 148: Final Spurt
Friday, June 29, Day 23

You put a fence behind me and in front of me, and you have placed your hand on me

(Psalm 139:5 EHV).

As the guys strolled along together, chatting throughout the afternoon, Carla and I hiked ahead. We began another long, gradual ascent. It was taxing, but according to the guidebook, a minor reward for our labors was ahead. The scenic overlook boasted views of Bear Mountain Bridge to the left and West Point to the right. Carla and I stepped out on the ledge, eager to catch our breath and admire a stunning view. All we saw was smoggy haze.

"Wouldn't it be nice to see a view?"

"Mom, who cares about a view? Let's get to the monastery."

"Right. It's only 4 p.m., and the real reward is only one-and-a-half miles ahead." I trotted back to the main trail with Carla scampering behind.

At the intersection of US 9 with NY 403, the traffic was so heavy we couldn't cross. Pickups, SUVs, Jeeps, and motorcycles whizzed past. Finally, with a tiny space between cars, Carla and I held hands, risked becoming fatalities rather than miss supper at Graymoor, and dashed across.

"Hey, way to go," Tom yelled from behind.

I grinned back at Tom and Curt, surprised they had caught us.

"My threat about lentils and rice must have worked."

The guys were stuck, waiting for traffic, and we weren't going to wait. Carla cupped her hands and yelled, "See you at the monastery."

Before the monastery, we entered one more swampland and picked our way over wobbly wooden walkways.

It was almost funny how on this last day of hiking, the AT had thrown everything at us, including running out of water, toe-stuffer descents, breath-gasping ascents, and now a swampland.

The AT continued with another brief, steep ascent, a grassy area, a paved road, a gravel road, a footpath through the woods, and an overgrown field. With relief, we glimpsed the orange, tin-can blazes that led us to Graymoor Monastery.

My eyes misted. I sniffed. The long, hard hike was over. God had kept us safe from behind and pulled us forward. We had arrived. At least, Carla and I had.

Vignette 149: Awesome Arrival
Friday, June 29, Day 23

He fulfills the desire of those who fear him; he also hears their cry
and saves them. The Lord preserves all who love him

(Psalm 145:19, 20 ESV).

Graymoor did not look like pictures of European monasteries, but rather a scattering of buildings set among winding paths and roads.

It was 5 p.m. when Father Cuthbert, a seventy-nine-year-old Irish monk, met us at the door.

"Yes, yes, three boys have indeed arrived, probably more than an hour ago." Father Cuthbert checked his watch. Behind him, the threesome appeared, beaming happily, freshly showered, and wearing relatively clean clothes. They danced and hooted.

"Yahoo!" Carla shouted as we bumped hips and joined in a jig.

Father Cuthbert shushed us and herded us down the hall to our rooms, one for each of us. They were simple with a single bed and a sink.

"Look, I have a Dixie cup dispenser." Carla dropped her backpack and ran to the sink for a drink.

I peeked into the next room. "I have a wooden, straight-back chair." I threw my pack on the floor and plopped down on the chair. "Ahh," I said, flattening my back. To us, the rooms were luxurious.

Shortly, Curt and Tom sauntered down the hall, escorted by Father Cuthbert, and the dancing and whooping continued. Behind them came Mountain Mist and an again-smiling Olive Oyle and Brutus.

Supper was a simple fare of fish cakes, baked potatoes, broccoli, and pudding, plus a beverage dispenser with unlimited lemonade and orange drink. Yes, unlimited. Father Cuthbert hovered over the kids, bringing extra platters until they were full, which took several platters.

After supper, we gathered with the other four hikers on padded chairs and couches in the library to share stories.

"Where were you folks three weeks ago during the big storm?" Brutus asked. "Wasn't that terrifying?"

"That was the second day of our hike." I shook my head in disbelief. "We were on top of Stony Mountain with lightning strikes nearby and thunder shaking the ground."

"We huddled in our tents, wet and scared, not sure . . ." Tom paused in his storytelling and glanced out the window.

"Was that thunder?"

All of us peered out the big windows as a rumble of thunder rocked the darkening room. Wind kicked up dust and stirred black clouds as a thunderstorm roared over us. It pelted the area with rain and hail for hours.

Once more I was reminded of God's kind protection on this hike. We were safe, well-fed, warm, and dry. A clean bed, not a flimsy tent, beckoned to us.

~Distance hiked on day 23: 12 miles or so
~West Mountain Summit to Graymoor Monastery

Vignette 150: Shaky Plan
Saturday, June 30, Day 24

Many are the wonders you have done, O Lord my God. No one can explain to you all your thoughts for us. If I try to speak and tell about them, they are too many to count
(Psalm 40:5 EHV).

We had a dilemma. How did we get our car 200-plus miles away? We asked around during a scrambled-egg breakfast.

"There's a bus stop near the bottom of the hill," Father Cuthbert said. "The bus goes into New York City and from there you can catch a bus or train to Harrisburg."

We formed a plan. Tom would go and the rest of us would wait.

Money was another dilemma. Tom had $9 in his wallet, and I had $15 that I'd planned for a donation to the monastery.

"Why so glum?" Mountain Mist saw our money piled on the table.

I glanced at his concerned face. "I have a checkbook and travelers' checks that only I can sign, but we have no way to cash either." I pointed to the table. "This is all we have for cash."

"Maybe this will help." Mountain Mist pulled $5 from his pocket. I'll give you my $5 donation, and you write a check for $20 to the monastery. That gives Tom $29."

"That does help." Tom smiled with relief. "I have a credit card, which may or may not work anywhere."

Later when we arrived at the bottom of the hill, I studied the wet, gravel turn-off. "Is this the right spot?"

"Well, he's waiting." Tom tipped his head toward a tattered man, duffel bag in tow, looking like an ex-con.

I nodded and chewed on my bottom lip. "What if your money runs out?" I feared Tom would get no further than NYC, and we'd become nuns and monks before he returned.

"I'll be fine." Tom patted my back, his eyes sparkling for the adventure ahead.

We turned and squinted down the highway. The bus was late. We waited longer. A Leprechaun bus whizzed by.

"Oh, no. Was that our bus? Did we not signal it? Were we standing at the wrong place?" My voice quavered in panic.

"This is a waste of time." Ben kicked the gravel.

"Do you have a better plan?" Tom clipped his words.

Our first plan was disintegrating, and so were we.

We waited ten more minutes. The ex-con character grabbed his duffel and yanked it up the hill. Tom and I glanced at each other. Now what? But before we could turn to Plan B, which was nonexistent, an old bus belching blue smoke roared to a stop in front of us.

"Do you take travelers checks?" I peered into the bus as the door swung open. The driver nodded. While I was fumbling with the checks, the ex-con ran back and boarded. I signed a couple more checks for Tom just in case he could find someone to cash them and hoped the ex-con wasn't hijacking the bus. Tom hugged the kids and kissed me good-bye. And the bus rattled off.

The five of us stood in the dust and smoke, watching it disappear around the curve at the bottom of the hill.

Vignette 151: Some Sweat, Some Don't
Saturday, June 30, Day 24

The Lord will watch over your going and your coming from now to eternity

(Psalm 121:8 EHV).

After the kids settled themselves in the library, I searched for Friar Paul, finding him at his desk. I needed to request permission for a second night at the monastery. Friar Paul leaned back, tenting his fingers in an ecumenical gesture, studying me.

"I'll gladly pay for our extra night." I paused. "How much is it?"

"You're small people and probably don't eat much." He drummed his fingers on the desk.

I held my breath. Obviously, he hadn't seen us in the cafeteria last night or at breakfast, devouring everything within sight.

"We charge $8 per person per night." His drumming fingers moved to his cheek. "You may stay an extra night and give whatever you think is appropriate." I wrote out an appropriate check, adding Mountain Mist's five dollars.

The kids were still content reading in the library, so I settled outside on a park bench to write in my journal and bask in the luxury of doing nothing. Not hiking, not carrying a pack, not answering questions, not setting up meals, not worrying who was ahead and who was behind me.

And that was our morning.

After lunch, the kids found a TV with limited channels, and that was our afternoon. What a restful day. Our only schedule was to arrive at meals on time.

Early evening, Father Cuthbert found us and handed me a piece of paper with a phone number scribbled on it. "Your husband called and wants you to call back."

My heart did a flip. Where was Tom? I retrieved my calling card and followed Father Cuthbert to the receptionist's phone. With shaky fingers I dialed.

When Tom answered, I blurted, "Tom, where are you? Is everything okay?"

"I'm in Harrisburg, but you won't believe my day."

"Did the bus break down?"

"No, I arrived in New York City with no problems. I could even use my credit card to purchase a Greyhound bus ticket to Harrisburg. Once on board, I settled back on the seat behind the driver, enjoying the ride, but on the outskirts of Philadelphia, the bus driver turned around and yelled, 'Does anyone know where the terminal is in Philly?'"

"You mean the driver didn't know where he was going?"

"Right. I helped him with the map. Near the terminal, angry, shouting people lined the street. Some threw rocks at the bus. When one hit the window by me, I ducked, but the driver didn't slow down, just plowed through the crowd."

I pictured a war zone. "What are you talking about?"

"Greyhound drivers are on strike, and my driver was filling in during the strike. People have even been killed during this vicious strike. Of course, now everything made sense—a driver who didn't know where he was going and a mob throwing rocks at us."

"Sorry your day was so stressful. We had a perfect, peaceful day here."

"I had a perfectly exciting adventure." Tom chuckled, a hint of relief in his voice. "I have our car and will be staying with the pastor's friend. See you about noon."

Vignette 152: Enlightened
Saturday, June 30, Day 24

In peace I will both lie down and sleep; for you alone, O Lord, make me dwell in safety

(Psalm 4:8 ESV).

I found the kids in the TV room, but not watching TV. Instead, Brother Larry held them spellbound with his stories about Brazil. He had spent twenty years there, building houses in poverty-stricken areas. The kids' eyes were big as Brother Larry described how grateful the people were for a new, one-room, tin-roofed shack—no electricity, no running water. I listened, glancing at the kids. Recognition lit their eyes. They knew exactly what Brother Larry was describing. They had lived it for a few long weeks on the AT. They'd slept in shacks and walked to distant springs for water. Once again I was reminded of the life lessons we would all take away from this hike, but it wasn't just the challenges, it was the people and their stories that made the AT unforgettable.

Brother Larry paused to catch his breath, smiled at his audience, and excused himself. Soon he returned with a soda and a glass of ice for each of us.

I reached for my frosty glass and can of 7-Up. "Thank you so much. How did you know the new love of my life? Every day I longed for a soda."

Brother Larry grinned. "I lived in Brazil. Remember? I know these things."

After a day off the trail, we could sip, not gulp, our icy sodas, so in between sips, I mentioned another wish. "We have been trying to see the lights of New York City for several days, at every lookout, but always smog blocked our view."

Brother Larry nodded. "I may be able to help you out with that too. The lights of NYC are indeed visible from our rooftop deck. Follow me."

We bounded up the stairs to the deck, excited to finally see NYC, but once again the smog won.

"I guess one fulfilled wish out of two isn't bad." Brother Larry nodded. "And with that, I bid you a peaceful night of sleep."

Back at our rooms, Ben discovered that he'd locked himself out. The receptionist called Father Cuthbert to come to our rescue. Again. He shuffled down the hall, stopping in front of Ben's room. Working through his jumble of keys, Father Cuthbert tried several until he was successful. With a kind *good night,* he shuffled off.

Even though we'd been shown nothing but kindness, I imagined a few old stories of rapes and murders in monasteries, so I showed the kids how to lock their doors securely from the inside. Without Tom and the other friendly hikers there, I felt alone and maybe a little vulnerable. I turned to the kids. "I'm just in the next room or across the hall. Don't open up the door to anyone. If someone knocks, yell for me."

They nodded solemnly.

Of course, Carla sensed my fears. "Mom, may I sleep in your room? I'll just roll out my sleeping bag on the floor."

"Nothing would make me happier."

Following Brother Larry's advice, we all had a peaceful night.

Vignette 153: Wheels
Sunday, July 1, Day 25

For your steadfast love is great above the heavens; your faithfulness reaches to the clouds

(Psalm 108:4 ESV).

During Sunday morning at the monastery, we fumbled around for things to do. I needed to move on, but we were stuck there until Tom arrived.

Finally, when we were in line for lunchtime salads, Tom walked in, all smiles. We were giddy with excitement, all but Ben. "Dad, what took you so long?"

"What took me so long?" Tom grabbed Ben by the shoulders. "I drove over 200 miles this morning and am here in time for lunch." Tom wrestled his way in front of Ben in the salad line.

"Okay, okay!" Ben stepped back with a sheepish grin.

"It was like flying." Tom continued. "I whizzed along all the mountains that we'd labored over and around. Trees flew past my windows in a blur. I love our car."

After a lunch and good-byes to Father Cuthbert and Brother Larry, we scampered out to our beloved brown Pontiac station wagon.

I patted its smooth panels. "I'm thrilled to be at this moment, done hiking, climbing into a car. Now how should we repack?"

We didn't really repack, just dumped the contents of all the packs but mine into the back of the station wagon. Nothing was easy to find, but accessible. We hoisted the empty packs into Kowalskis' car-top carrier. My pack held the first aid kit and other essentials, so we kept it intact and tossed it into the back of the station wagon too.

With a sigh, I settled into the front seat. Oh, the joy of sitting on a padded high-back seat and watching the world pass effortlessly by.

We explored West Point, gawking at the chapel's beauty, and circled the campus several times, trying to find the lookout over the Hudson.

"We've seen the education building four times," Ben whined.

Already our joy at being in a car was diminishing as six people crammed into six seats, elbow-to-elbow. I was feeling the rumble of a revolt in the backseat. "Let's just stop at the visitors' center for information on camping and a grocery store."

Later, Tom shook his head as he maneuvered our car up Bear Mountain in Harriman State Park, where the closest campground was located. "Can you believe that two days ago we hiked through this area, desperate for water and dreading the eight miles to hike to the monastery?"

But as we drove around and around the park, unable to find our camping spot, we longed for the simplicity of the trail. And when we finally found our tiny site, monstrous RVs towered on each side. We squeezed our tents between the RVs and looked around with disgust. Litter and filth were scattered everywhere.

But one wish did come true. After dark, we drove south on the Palisades Highway until we came to a lookout across the Hudson. There, in all her glory, the Big Apple strutted her zillions of lights.

Vignette 154: Peaceful Endings
Monday, July 2, Day 26

Return, O my soul, to your rest; for the Lord has dealt bountifully with you. For you have delivered my soul from death, my eyes from tears, my feet from stumbling; I will walk before the Lord in the land of the living

(Psalm 116:7-9 ESV).

I studied Tom across the picnic table. His face was sweaty, his eyes bleary. The hot air vent from our neighbor's RV blew across my back and blasted Tom in the face. Harriman State Park was still asleep.

"It's peaceful this morning." I squinted into the early morning sun.

Tom snarled. "Of course it's quiet. The campground was rocking half the night. I can hardly wait to get out of here."

"Me too. Let's hit the road."

Our mission for the day was to pick up mail that had been sent farther up the trail. Our first stop was Fort Montgomery to collect letters from friends plus a food box at the post office.

We continued, paralleling the AT. At Peekskill, we felt the weight of being surrounded by little houses and crowded streets. There we redeemed buy-one-get-one-free coupons at McDonalds, moving further away from life on the AT.

Another food box awaited us in Kent, Connecticut, and more letters and boxes from friends. We read our letters aloud as we drove to the post office in Tyringham, Massachusetts. No mail there.

A local lady directed us to a camp at Beartown State Park. All twelve sites were filled for the night, so we drove down to a quiet part of the lake to relax. It was a pristine lake surrounded by old-growth trees.

While Tom and Carla cooked chicken noodle soup and pistachio pudding, I repacked our food, keeping out only what I thought we'd need during the next few days. I tossed the rest into our now-overloaded car-top carrier.

The boys fished, and Carla waded near the edge, trying to catch minnows. At dusk, Carla and I walked on a country lane, following arrows to the AT, enticed to see it once more.

When we returned, I announced, "I miss the AT." The Appalachian Trail disease had claimed me, the desire to hike it, to go further, always with me.

"We don't." It was a unanimous family exclamation. The Appalachian Trail disease had nibbled at the others but never claimed them.

That night we were too tired to move our car to another noisy campsite.

"This place is perfect." Tom pointed beyond the car to a patch of grass sheltered by a tree canopy. "Let's set our tents there and hope nobody will move us along."

The last night in the wilderness for Swiss Family Wisconsin was peaceful, no noise but the singing of the breeze through the trees and the lapping of waves on the shore. Tomorrow we'd begin our freeway dash home.

~Distance hiked from day one through day 23: 262 miles

~ Pennsylvania Route 225 to Graymoor Monastery, New York

Where Are We Now?

The memories of hiking the Appalachian Trail hold onto each of us. It was, above all, a great family adventure. We bonded through the struggles and laughter and we learned the value of taking one more step—on the trail and in life. Did we emerge at the other end in super physical shape? Yes. Stamina and flexing our thigh muscles come naturally. But, most importantly, we learned the lesson of Psalm 4:8:

> In peace I will both lie down and sleep; for you alone, O Lord, make me dwell in safety.

Tom taught at Winnebago Lutheran Academy, a high school, until 1995 when he went into business for himself as the owner of Tom's Christian Tours. Every time he traveled by motor coach to the East Coast, he'd watch for signs of the AT crossings and launch into a tale of his adventures for the group. He never did another long-distance hike. Atrial fibrillation has continued to plague him, but his knees recovered.

Janie wrote for a small newspaper, worked as a teachers' assistant, authored four children's books, and joined Tom in running their group travel business. When she saw signs for the AT, she longed to return to it, but hasn't. Instead, she has hiked long sections of the Pacific Crest Trail with friends. Backpacking is fun, never easy, but sharing the camp chores with other moms is a breeze.

Ben continued on his road to physical fitness, graduating from college with a degree in personal training. After a few years, he changed career directions to become a financial advisor. Ben and his wife Sara live in Colorado with their three cats and two dogs. They often hike together, including climbing to the top of some mountains over 14,000 feet.

Caleb's love of fishing followed him into adulthood. As a high school physical education teacher, he wrote a grant to make fishing a part of his high school curriculum. Caleb also heads his foundation, Fishing for the Heart. His goal is to teach under-privileged kids to fish, all in a Christian atmosphere. Caleb and his wife Jill have two children, Livie and Graham.

Joel, while attending Georgetown University, organized his friends to hike a section of the AT, but they preferred going out to dinner more. Joel moved to New York City in 2000, enjoying the glitz of the city we longed to see from the trail. He has continued his flare for technical/computer work, often working as an independent consultant. Recently, Joel has made Sao Paulo, Brazil, his home.

Carla traveled with the multi-cultural group Up With People, performing in the USA, Mexico, and Japan. She returned home to marry her high school sweetheart Dan. The Air Force took them to North Carolina, Alaska, and Georgia before they settled back in Wisconsin. They have four children: Avery, Alix, Bryce, and Abby. Their second child, Alix Ann, was born in Alaska exactly seventeen years after we began the AT hike. Carla owns a photography business.

Front from left: Joel Niedfeldt; Graham and Livie Niedfeldt; Alix, Bryce, and Abby Jahnke
Back from left: Sara and Ben Niedfeldt; Jill and Caleb Niedfeldt; Janie Niedfeldt; Dan, Carla and Avery Jahnke; Tom Niedfeldt

What Did We Carry In Our Six Backpacks? Too Much!

Pocket Bible and devotional book; three tents; plastic ground cover for three tents; one tarp, duct tape; six sleeping bags; two sleeping pads; ten canteens; water purifying pump; rope and clothes pins; backpacking camp stove; twelve canisters of fuel, matches in waterproof containers; a big pot, medium kettle, and skillet; one cooking spoon; one spatula; campfire popcorn popper; six plastic plates; six knives, forks, and spoons; six plastic cups; condiments like salt, pepper, sugar, cooking oil, powdered milk; food for first week; two changes of clothes each; windbreaker and fleece for each; swimming suits; gloves & stocking hats; camp shoes; rain coats; mosquito netting hats; mosquito repellant; several books; big camera with extra lenses; binoculars; harmonica; extensive first aid kit and vitamins; sewing kit; toiletries for six; toilet paper; six lightweight towels and washcloths; trail guides & maps; notebooks and pens; playing cards; assorted fishing gear; six flashlights & extra batteries

CPSIA information can be obtained
at www.ICGtesting.com
Printed in the USA
JSHW021241160520
5721JS00003B/3